CW00431915

THE NEW NOAH
A CLIMATE CHANGE SURVIVAL GUIDE

THE NEW NOAH
A CLIMATE CHANGE SURVIVAL GUIDE

Peter Dawe

BANK HOUSE BOOKS

First published in the United Kingdom in 2008 by
Bank House Books
BIC House
1 Christopher Road
East Grinstead
West Sussex RH19 3BT

BANK HOUSE BOOKS is a division of BANK HOUSE MEDIA Ltd

© Peter Dawe, 2008
Additional text by David Long

All rights reserved. No part of this publication may be reproduced, stored in a retrieval system, or transmitted, in any form or by any means, electronic, mechanical, photocopying, recording or otherwise, without the prior permission of the publisher and copyright holder.

British Library Cataloguing in Publication Data
A catalogue record for this book is available from the British Library.

ISBN 9781904408352

Cover illustration by Sue Randle after a fifteenth-century engraving of Noah's Ark.
Lightning photograph by Jeannette Sutherland

Typesetting and origination by Bank House Books
Printed and bound by Lightning Source.

CONTENTS

	Introduction	1
Chapter 1	The Climate Crisis	5
Chapter 2	Too Late: A Failure of Response	19
Chapter 3	The Futility of Gesture	30
Chapter 4	Carbon Capture and Sequestration	42
Chapter 5	The Carbon Offset Con	55
Chapter 6	The End of Cheap Energy	64
Chapter 7	Saving Energy Won't Save Us	76
Chapter 8	The Biofuels Fallacy	90
Chapter 9	Food as Fuel	103
Chapter 10	The Impact of Population	113
Chapter 11	Migration and Meltdown	129
Chapter 12	A New Noah	144
Chapter 13	Find Your Tribe	157
Chapter 14	Knowledge, Trade and Culture	164
Chapter 15	The Wash Tidal Barrier	178

INTRODUCTION

❖ Climate change is happening

❖ It's already too late to stop it

❖ The impacts will be widespread and life-changing

❖ Here's what *you* can do to preserve what you value most

It is good that people are at last showing some concern about the future, and not just because – as the scientist and social philosopher Charles F. Kettering once put it – 'we will have to spend the rest of our lives there'. But unfortunately that concern is not strong enough, and it comes too late in the day to make enough of a difference: the truth is, the time for talking is already over.

Instead, global climate change has become a reality, with even the likes of the United Nations accepting that levels of atmospheric carbon dioxide have already passed danger level. In only one sense does the option to do something about it still exist: if we stopped digging and drilling fossil fuels out of the ground now these same levels would slowly begin to fall. But in a world built on cheap and plentiful energy this is never going to happen, nor are schemes to make homes, corporations and even entire cities 'carbon neutral' ever going to achieve the drastic cuts in emissions which are necessary if the world is to avert disaster.

As a consequence we need to plan for the future rather than seeking to change it. To accept that climate change is going to impact heavily on how and where we live our lives. To conceive of ways in which we as individuals might safeguard what we most value, and – with the help of this book – take steps to achieve what we can in this regard before it is too late.

Of course the existence of climate change itself, and of the greenhouse effect, was never in any doubt: without it the temperature of the Earth would be a frozen minus 18°C instead of its present, more comfortable 14°C average. Even so, the whole subject remained until very recently highly controversial, especially when it came to evaluating and agreeing whether or not the principal cause of global warming was human activity.

Ten or fifteen years ago climate change was mostly a matter of conjecture, with expert opinion strongly divided on the pace of change, and what the impact of those changes on our lives were likely to be. Today, though, the argument is more about the speed and the scale of the changes rather than whether or not they are happening, and the debate is about the best ways to avert the impact these changes will have on our lives, rather than whether or not there will be any such impact.

But, as this book demonstrates, the debate is already a stale one; the truth is that the future is unfolding before our eyes, or maybe even unravelling. Even this, however, has not been sufficient to stop people talking in general terms about 'saving the planet' when what is really at stake is not the Earth – which as a big rock will likely continue circling round the sun for the foreseeable future – but rather our lives upon its surface and the long-established ways in which we choose to live them. As a result very little is being done at any level to address the changes which are already making themselves felt, and more significantly to ascertain what might be the best responses to these changes.

Because of this avoidance and delay it is almost certainly too late to make the changes which might provide a route around the disaster. For one thing the political will is lacking, as indeed are the vision, the mechanisms and the global consensus which would be required to slash carbon emissions to levels that might make a difference. Instead, and at almost every level of society, people remain resistant to the fact that as long as we go on drilling and digging carbon out of the ground it will find its way into the atmosphere, and temperatures and sea levels will continue rising.

That's why this book has been written. To demonstrate the scale and immediacy of the problem, and the futility of our responses so far. To warn individuals and organisations of the dangers of grasping at straws every time they read or hear something which appears even remotely positive. To underline the fact that coping with the effects and aftermath of climate change is a global challenge, even though many of the examples in these pages may seem rather small or

localised. And finally to expose the 'greenwash' which accounts
for so many of the so-called solutions that have been proposed
over the last few years, and which continue to be promoted by
businesses, governments and international organisations
around the world.

Most importantly, though, this book has been written
to identify ways in which each of us can take action to help
ourselves. Not to avert disaster – something which is certainly
beyond the ingenuity and effort of any private individual – but
rather to salvage from it what we personally most value. And
take action we must, because we face a singularly severe threat.
Not in terms of our outright survival, perhaps, as mankind is
sure to survive, even if in diminished numbers, in reduced
circumstances or in fewer geographical locations; but certainly
when it comes to ways in which we choose to live our lives and
raise our families.

By assembling the key facts and opinions, this book
attempts to show that even though the predictions of doom
may be out by an order of magnitude – we may face disaster in
only ten or fifteen years, or it could be fifty or seventy-five;
similarly the scale of the problem may be a few mass
migrations from low-lying areas or the depopulation of entire
continents – what we cannot do is assume that everything will
be all right, or that we can simply carry on as we are now.

Instead, the big philosophical problem is deciding
what we want to preserve and then establishing practical,
workable ways in which we can do that. Do we wish to
safeguard ourselves, our families, individual nations or all
humanity? Is it society as a whole which we value, or a number
of different societies? What about knowledge and culture? We
must even ask whether we are willing to make choices about
sacrificing other people, even entire nations, in order to ensure
the survival of our own.

Only then can we work out how to preserve what we
wish to save, and this book will show you how.

Chapter 1

THE CLIMATE CRISIS

Simply because something is natural or has happened before doesn't mean it won't have serious consequences.

The evidence for climate change is widespread and its impacts, while often strange and diverse, are no longer mysterious or hard to discern. Blips in the chart do not disprove its existence: the trend over time is upwards and sharply so. At one end of the spectrum new and different weather patterns are causing active desertification in much of the developing world. At the other, and as this chapter was being written, many millions of individuals were displaced by widespread flooding across Asia and central Africa, with several hundred already dead. Similarly, on the very day a study published in the US revealed that the number of hurricanes hitting the southern states had doubled in recent years, the highly respected journal *Nature* observed that the Arctic spring had arrived a full two weeks earlier than had been the norm ten years before.

Mounting evidence impossible to ignore
At both poles there is plenty of evidence of change, and of the quickening pace at which the once vast ice fields are disappearing. In September 2007 the extent of Arctic ice was 23 per cent smaller than the previous record low – and nearly 40 per cent less than the average annual minimum recorded between 1979 and 2000 – while the fabled North West Passage was open to shipping for the first time in recorded history. Similarly the extent of the annual snow cover across the whole of the northern hemisphere, having decreased by approximately one-tenth since 1966, has stubbornly remained below average every single year since 1987.

At lower latitudes too climate change is making itself felt with increasingly regular, increasingly lethal storms – and everything that comes in their wake, such as flooding, landslides, crop and power failures, hunger, diseases (particularly those associated with diarrhoea, a leading cause of death in the developing world), property damage, poverty and economic turmoil and of course loss of life. Meanwhile drought in parts of the US and in Australia (the worst since 1895–1903, when half of all livestock died of dehydration) has contributed to a global shortage of such staples as milk powder, a key ingredient in many processed foods, which has seen a price rise of almost 250 per cent in a single year. (In June 2007 Britain's *Daily Telegraph* reported that the price of milk powder had risen from £1,000 a tonne to £2,400, with most of the growth coming 'in the past few months'.)

Nor has equable, temperate Europe and continental America managed to escape unscathed. Glaciers to the north and at high altitudes are melting fast, and further south forest fires are becoming more frequent and far more destructive. We are also seeing a growing incidence of increasingly ferocious summer heatwaves: in 2003 one of these accounted for more than 35,000 deaths among Europe's sick and old, since when it has been established that the number of extremely hot days in Europe has nearly tripled in the last 120 years, while the duration of heatwaves over the same period has similarly doubled.

The pace of change is accelerating fast

That said, of course, changes of this sort – global warming as well as global cooling – have occurred many times in the past. There is also a body of reliable scientific evidence proving that periodic fluctuations of this sort can be entirely natural, demonstrating that many examples of such changes occurred in the years before man and man-made carbon emissions began to make themselves felt.

But simply because something is natural or has happened before doesn't mean that it won't have serious consequences or that the threat from it won't escalate as time goes on. This is especially true because the world's population is so large, and because so many millions of us live in homes and even entire cities built in places which until relatively recently (and with good reason) were only sparsely or intermittently inhabited. There is also plenty of evidence suggesting that the world has not been this warm for a thousand years or more, and that the pace of change has accelerated recently and continues to do so. It is known, for example, that while reliable temperature records go back only 150 years or so they all disclose the same pattern: that the twenty hottest years on record have all occurred since 1981; that the ten hottest have occurred since 1990; and that three of the four warmest years in the last century and a half have occurred since 2002.

Not just warmer but wetter

In recent years the Earth's atmosphere has been getting wetter too. This is a cause for concern since additional water vapour presents itself as yet another greenhouse gas, as well as fuelling the development of hurricanes. Furthermore, and as if such statistics did not themselves provide sufficient warning of what is happening to the planet, these changes have happened in spite of many natural occurrences over precisely the same period – changing solar cycles, for example, or the action of volcanoes that inject particles into the atmosphere, thus reflecting solar radiation back into space – which should have acted to cool things down.

To explain this we are left with the conclusion of the climatologists on the UN Inter-Governmental Panel on Climate Change. In their 2007 report, *The Physical Science Basis: a Summary for Policymakers*, IPCC authors agree and state clearly that this rapid warming effect is to a very large and measurable degree caused not by natural phenomena but by normal, everyday human activity of the sort in which we all partake. Not that anyone denies there are many different natural emitters of CO_2 to be taken into account – and, yes, it is true that CO_2 emissions due to human activity account for only a very tiny fraction of the global total of carbon emissions. But without allowing for human intervention it is difficult if not impossible to conceive why the level of atmospheric CO_2 is climbing at its current rate, or to explain the climatic changes which are now being observed.

CO_2 levels higher than at any time in 20 million years

One needs to consider too, for example, that for at least 500,000 years (according to US National Oceanic and Atmospheric Administration figures) the background level of

CO_2 consistently hovered around 280 parts per million by volume-. Then, quite suddenly and with the onset of the industrial age in Europe, the level began to climb steadily, to reach 455 ppm by 2005 according to United Nations expert Tim Flannery. Now it is rising even faster, at more than one and a half parts per million per year, taking the concentration of CO_2 in our atmosphere to a level which has not been seen for approximately twenty million years.

As a result, says another IPCC document, a *Special Report on Emission Scenarios* (2000), by the end of the century the concentration is likely to be higher still – somewhere between 490 and 1,260 ppm – which is to say up to 350 per cent higher than it was in pre-industrial times.

By any standards that is an unprecedented rate of change – about 10,000 years' worth compressed into fewer than 100 – and with more CO_2 in the air now than at any time since humans first learned to walk upright, we are beginning to discover just how poorly the planet can deal with such rapid changes. Slow changes appear to be all right – they allow time for individual species and the biosphere as a whole to adapt – but the sort of rapid changes we now see cannot be accommodated, and instead seem certain to cause biological chaos as well as massive, widespread disruption to many different aspects of agricultural production.

Forecasts vary – but none is optimistic
There are many reputable scientists working in the field who are confident that these particular impacts will be felt much, much sooner. Their assertions are based on the fact that when it comes to dramatic climate change the precise degree of how far and how fast is determined not just by the amount of carbon we produce but by many different feedbacks as well. Many of these are still much less well understood: melting ice, for example, the oceans, water vapour, clouds, changes to afforestation and so forth. Thus reports such as the one

prepared for the Climate Institute at the end of 2007 – by researchers at the University of Melbourne – drew attention to science emerging since the 2006 IPCC announcement, which suggested that temperatures and greenhouse-gas pollution are both rising much faster than the earlier report suggested.

In particular the Australian data demonstrated that the global-warming trend is accelerating faster than expected and would, if such a trend continued, lead to a temperature rise of approximately 3°C by the end of this century – this being relative to pre-industrial temperatures – thereby tipping us into the danger zone which is currently defined as any rise in excess of 2°C. The report found a similar acceleration in the growth of CO_2 emissions, increasing from 1.1 per cent per year for the period 1990–99 to more than 3 per cent per year during 2000–4. Such a rate far exceeds that in even the most fossil fuel-intensive emissions scenarios used by the IPCC, and indeed shortly afterwards the IPCC itself hardened its approach considerably.

Even without such considerations, however, the stark reality is that the window of opportunity for any smooth transition to the much-vaunted low carbon, clean energy economy we hear so much about these days is closing much faster than most forecasts appear to suggest. Indeed, by rapidly converting earth's once vast reserves of oil, coal and gas into literally billions of tonnes of CO_2 annually – and with the effect of this magnified by releases of many other planet-warming substances such as chlorofluorocarbons, water vapour and methane – we have already succeeded in trapping more and more solar radiation within the troposphere. Through a mechanism which is now well understood, the effect of this has been to raise average temperatures around the world by approximately half a degree with another 0.7°C in the pipeline – that is, ready to make itself felt thanks to the time lag which is inherent in the system.

Human impact no longer in doubt

In America, for example, the National Oceanic and Atmospheric Administration confidently asserts that human-produced greenhouse gases are responsible for more than half of this rise – for more than half of all warming which has been seen in the US in recent years. In 2006 average temperatures came to within half a degree of the highest average ever recorded in that country. It is significant too that the aforementioned half a degree rise has been made since 1970, that's up from a baseline temperature which has held steady since the first cities were built and the first quasi-domesticated crops were grown several thousand years ago.

Admittedly, moving forward, the precise extent of the increase we should expect and the timescale in which it will happen are both still open to conjecture. Taking a sensible median of the more obviously respectable studies, however, a rise of 2°C overall seems eminently feasible over the next few decades, and the likely effects of this (deceptively modest-sounding) rise are already well understood.

In the cryosphere, for example, many of the largest changes are already easy to observe and monitor, and very easy to measure. The Arctic ice – beneath which nuclear submarines still like to hide – has actually lost more than 40 per cent of its thickness since the height of the Cold War. Worse still, the rapid decreases in the extent of this ice are occurring much faster than climate model projections have so far forecast, with the current summer minima approximately thirty years ahead of a range of different simulations, prompting suggestions that a completely ice-free Arctic Ocean might occur much earlier than the previously suggested 2050.

Earlier still, by 2015, the celebrated Snows of Kilimanjaro will almost certainly need to be rebranded the Rocks of Kilimanjaro. (The transformation is already well under way, with 82 per cent of the white stuff having gone since the territory was first mapped in 1912.) And Montana's

stunningly beautiful and highly popular Glacier National Park
will similarly be more or less glacier-free within just two
decades, an alarmingly short timescale when one is still
inclined to think of any major planetary changes as occurring
in geological time and moving with so-called glacial speed.

Nature just can't keep pace

But of course climate change is not simply about the loss of
amenity or the destruction of cherished and beautiful natural
landscapes. Nor of course is it simply a matter of Britons
enjoying warmer if somewhat wetter summers, and lower
heating bills throughout the winter months.

Across the world this glacial melting already causes
rivers to flood at high altitudes, while lower down the slopes
faster evaporation means that others dry up, with devastating
consequences for those living on their banks. In the same way,
while in some parts of the world commercial crops are
beginning to ripen much faster than they should do, in many
other regions crop yields are collapsing dramatically as a result
of drought and disease.

Elsewhere, as more destructive types of hurricane increase
in both frequency and strength, fears are growing about the
possible failure of the mighty ocean currents which have so far
kept the European climate comfortable and stable. There is also
widespread concern that nature's so-called carbon sinks are failing
to grow in step with the increased levels of CO_2 being produced.

Examples of these include soil, vegetation and large
bodies of water, each of which is a perfect and elegant
illustration of how over many millions of years natural
processes evolve to achieve a careful balance between carbon
storage and its release, in order to maintain a global
environmental equilibrium. But unfortunately the capacity of
one of the most effective, the Antarctic Ocean, is now thought
not to have increased at all since the 1980s, despite CO_2
emissions having risen by around 40 per cent in the same time.

Other likely impacts are rather less predictable, or more commonly of less concern to the general public: for example, the numerous examples now seen of diseases spreading beyond their traditional geographical limits as pathogens and their hosts are enabled to migrate. Sometimes the lack of concern is simply because we exercise our option not to think about them, or because we lack the expertise or ability to comprehend the wide-ranging effects of, say, the reduced biodiversity that must by definition result if natural ecosystems such a coral reefs and rainforests are badly disrupted or even deliberately destroyed.

Of course it is also possible that many choose not to engage with the problem on the assumption that science will be able to produce the appropriate inoculation or cure and within an acceptable timeframe. This, however, is highly unlikely – particularly as more multi-drug-resistant strains evolve – and in fact there is clear potential for epidemiological transition to turn out to be one of the more severe side-effects of climate change.

Many non-scientists similarly find it hard to conceive how, for example, were we to precipitate an irreversible melting of the Greenland ice sheet the resulting meltwaters would be sufficient to raise sea levels by as much as 6 or 7 metres. Such a rise, which has been posited by Tim Lenton of the University of East Anglia, could also result if the warming of the Antarctic Ocean continues to destabilise the West Antarctica ice sheet. Were such a thing to happen it would permanently flood land currently occupied by literally hundreds of millions of people – and according to the IPCC's own figures flooding of this magnitude could be precipitated by a temperature rise of just 1°C.

Impacts differ – but no-one will escape the storm
Obviously different countries will suffer to varying degrees, but there is little doubt that even were sea levels to rise by a

more modest amount, among the hardest hit would be the heavily populated areas around the great Asian deltas, while many smaller, marginal or island communities could disappear altogether.

Indonesia alone estimates it could lose around 2,000 islands by 2030, a scenario described by Environment Minister Rachmat Witoelar but which only ten or fifteen years ago would have been dismissed as too far fetched even as a plot for a disaster movie. But today the science of global warming is much better understood, and while in richer countries such as the UK many continue to dismiss all such concerns on the grounds that, for example, slightly warmer summers would be a welcome change in northern Europe, the reality is that even a very modest rise in average temperatures (the IPCC's own estimates vary from 1.4 to 5.8°C) would wreak havoc across the entire globe.

Were, for example, average temperatures to rise by a seemingly modest-sounding 3°C, giving northern latitudes a climate like that of south-west France, one has to ask where all the Spanish will be living by then, and the French, and all the people struggling to survive in North Africa.

Or for that matter all the people in the southern US, another area already becoming drier and hotter and likely to become uninhabitable desert.

It's not just that these and other locations would become less comfortable and less productive, but that a good deal of currently heavily populated and highly productive coastal land would disappear completely.

The IPCC report, for example, forecasts sea-level rises of around 0.6 metres – which is bad enough when you start measuring the altitude of many tens of thousands of square miles of coastal land. But their estimate was made without allowing for a number of key variables, including changes in sea ice, clouds, water vapour and the effects of aerosols; and anyway, the last time average temperatures were this high (which was three million years ago) sea levels were, says NASA physicist Hames Hansen, nearly 25 metres higher than today rather than 0.6.

Indeed, having studied and modelled the human impact on climate since the 1970s, Hansen says such a rise is not just feasible but likely, and as such would be sufficient to flood the homes of at least 50 million people in the US, many of them on the east coast. Around 250 million would similarly suffer in China, together with 150 million in India, and in Bangladesh another 120 million people, the latter being roughly equivalent to that country's entire population.

Flooding a first-world problem too

It's worth considering too the *New Scientist*'s assertion (28 July 2007) that some 125,000 years ago, when average temperatures were a mere 1–2°C warmer than they are now, sea levels nevertheless rose by 5–8 metres. In today's context even the lower end of this range would be sufficient to force the total evacuation of London, New York, Sydney, Mumbai, Tokyo and countless other major centres of population, while leaving their surrounding areas highly vulnerable to storm surges. It's a sobering thought too when one considers that, whereas over the last 3,000 years sea levels have risen at an average rate of 0.1 to 0.2 millimetres a year, tidal data indicates that we are experiencing a rate of roughly ten times this already. Such figures also underscore once again the fact that these climatic changes will eventually affect us all. Not just the poor in developing nations, but – as the OECD itself recognised in a report published in late 2007 – everyone in the West as well, as storm surges and floods made worse by global warming begin to threaten literally trillions of dollars-worth of buildings, property and infrastructure. According to the OECD Miami finds itself at the top of the threat-list, with $3.5 trillion of 'exposed assets', but it is very closely followed by Guangzhou in China ($3.3 trillion) and New York with $2.1 trillion.

Even without this major threat of complete
inundation, however, these same relatively small changes in
temperature are already beginning dramatically to alter the
biological and agricultural profiles of the oceans and virtually
every landmass which surrounds them. This can be seen, for
example, in the shifting patterns of migrating cod stocks (and
the economic impact of this) and in the way in which malaria
is spreading beyond its traditional boundaries. Bringing with it
an obvious impact on social and health issues, the spread of
malaria is overburdening already strained health services in a
pattern likely to be repeated many times over.

Similarly, in the northern hemisphere, plant and
animal species which can migrate will gradually do so, moving
over several generations in order to colonise cooler climate
zones further north. In a sense this move has already started,
with several big French wine growers acquiring land in
southern England, a region where the quality of English wine
is fast improving – although of course this too could be
negatively affected by changes to the local climate. North
American wheat production might similarly shift from the
mid-west up into Canada, since with two-thirds of global
water consumption already being used in agriculture farmers
further south are finally beginning to recognise that they will
not be able to simply irrigate their way out of the problem.

Unfortunately, of course, looked at in the round few
things are ever this straightforward or as easy to accommodate,
and many new climate areas will quickly prove themselves to
be wholly unsuitable for large-scale agricultural activity of the
sort on which we depend. Areas which were recently glacial,
for example, may inherit the right temperature profiles over
time, but they will still be characterised by bare, scoured rock
and scree or, in the best scenario, by thin, new soils lacking in
the necessary levels of nutrients. These will be nothing like the
rich soils we actually need, and even given the natural
resources required to boost their fertility artificially – a move

with its own very serious environmental costs and implications – it would take many years for them to develop and mature to a point where the areas in question can even come close to matching the depth and fertility of the soils in the more traditional agricultural areas of Europe, North America and Asia.

More mouths, less food

The likelihood is that agricultural yields will fall in almost every country of the world, that both subsistence and commercial agriculture will suffer very badly as a result, and that between a quarter and a third of all plant and animal species will face imminent extinction. Rice fertility, to take just one key example, is known to fall by 10 per cent with every one degree rise in temperature, and to reduce to zero once the temperature rises above 40°C. That's certainly food for thought given that rice is the world's single most important food crop, supplying around three billion individuals with more than 80 per cent of their diet.

Moreover, in the worst-case scenario, this rise to 40°C and consequent drop off in rice production could happen long before the end of the century. In fact many climatologists think it could happen much sooner, particularly if, as seems increasingly likely, the major ice sheets suffer a catastrophic collapse. It goes without saying that many communities will suffer far more and far more immediately than we will in the West – for example farmers and their dependents in Africa, where many experts fear that crop yields, already way below western levels, far more sensitive to change and therefore highly fragile, could actually halve as soon as 2020.

The impact of this on the local population will naturally be devastating, although the truth is no country will escape the fallout entirely. Not because we in the UK will be robbed of the opportunity to eat cheap, air freighted green beans grown out of season and flown in from Kenya – though what we are actually doing here, incidentally, is importing water from a country with relatively little of it, anything up to 5,000 tonnes of it per tonne of crops imported – but because in a complex and increasingly fluid global economy like the one we have now the domino effect of this kind of disaster is absolutely guaranteed.

Chapter 2

TOO LATE: A FAILURE OF RESPONSE

Many of us feel that nothing we can do on a personal scale
can possibly make a significant dent in such a massive,
worldwide problem – and in this we are, for the most part,
almost certainly correct.

The requirement to make major cuts in CO_2 emissions is now widely understood and generally acknowledged by scientists, governing authorities and the public – yet the efforts to do much about it have so far accounted for very little.

This is true even in the richest developed nations, so that in the US emissions of CO_2 have actually risen, by around 15 per cent since 1990. (California's record is equally poor in this regard, despite the energetic campaigning and impassioned words of its famous Hollywood 'Governator'.) Emissions in Japan have similarly increased by more than 10 per cent, and although the member states of the expanded European Union can together claim to have cut their CO_2 output by 2.9 per cent, much of this reduction has actually resulted from the transfer to the developing world of manufacturing and heavy industry rather than from any specific environmental initiatives. Unfortunately global emissions of other greenhouse gases, such as methane, have risen over this period.

It's not our fault, but theirs – or is it?

Inevitably a certain amount of buck-passing has also served to obscure the problem. For example, and in part as a consequence of television images of opaque clouds of pollution hanging over its new Olympic Village, it seems now to be generally accepted throughout Europe and the US that China poses the biggest threat. Refusing to listen to reason, goes the argument, the authorities there are busy opening a new, smoky, coal-fired power station every four days, and planning to build around seventy major airports over the next few years. Little wonder that the latest forecasts from the International Energy Agency suggest that by 2009 China will have overtaken the US to become the world's biggest single polluter.

But the reality is somewhat different. On a *per capita* basis the developed world is still the biggest sinner and by a country mile. The gap may be closing but in 2003, while the Chinese and fast-industrialising India emitted 3.2 and 1.19 tonnes of carbon per head respectively, the average American, German and Briton still accounted for 19.8, 9.8 and 9.4 tonnes apiece. Even the French, who obtain a much higher

proportion of their energy than we do in the UK using cleaner
nuclear power, dwarf China and India's head-for-head output
with an average of 6.4 tonnes. And let's not forget either that a
substantial proportion of China's CO_2 emissions are these days
produced making goods for western consumers rather than
their own. . . .

Opinions shift, but talk is not enough

It is at least true that in the developed world recently both
government and public bodies have been making more
positive noises about the problem – even if, all too often, their
actual responses have been marginal or poorly conceived. It is
true as well that big business and many of the major
corporations have woken up to the commercial challenges of
establishing more ethical and sustainable business practices –
in December 2007, 150 of the world's biggest corporations
including Shell and Nestlé called for tougher targets to cut
emissions by 50 per cent by 2050 – although across the
business sector as a whole the majority of operators admitted
as recently as June 2007 that it was not yet a high priority for
them. (According to the *Financial Times* report of a KPMG
survey, climate change was still bottom of the priority list for
Britain's largest companies, only 14 per cent of which had a
clear strategy for dealing with it in the future.)

Others have meanwhile been merely tinkering at the
edges, and in some cases cynically devising new ways in which
to enhance their profiles and profits by appearing to be greener
than they actually are. As for private individuals, many of us
feel that nothing we can do on a personal scale can possibly
make a significant dent in such a massive, worldwide
problem – and in this we are, for the most part, almost
certainly correct.

In this regard, switching off a lightbulb or two
certainly isn't going to help, nor as we shall see (and
notwithstanding their growing popularity) will changing to

more elaborate low-energy ones. Many consumers also rightly suspect that if they don't use the oil someone somewhere else will, a factor which may help explain why – despite demonstrable increased public awareness of 'green' issues – sales of off-roaders and other so-called gas-guzzlers continue to rise, and in the UK actually reached a four-year high as this chapter was being drafted.

A further problem arises because when they discuss the impact and implications of climate change both individuals and politicians are in the habit of talking about future impacts and the implications of climate change for the future. This robs the problem of its urgency, as well as encouraging a sense that there is time to stop and think about how we might avert the looming crisis, when in reality very much the reverse is true. Global warming is already happening, and we are already witnessing some of its genuinely catastrophic effects. Perhaps most worrying of all, however, the generally localised changes and resulting disasters which we have observed relate to historic and much lower levels of CO_2 emissions rather than to the much higher levels being released today.

Why cuts aren't a cure

Just as there is a substantial and well-understood time-lag between the toxic outpourings of industrial chimneys and acid rainfall, between radioactivity and specific cancer strains, between soil depletion and human starvation, and between deforestation and drought, climatologists now recognise that a considerable delay occurs between the escalation of greenhouse gas levels and the impact these have on climate.

What this means, of course, is that – given the much larger volumes of greenhouse gases which are being released today – when it comes to climate change and its consequences there is clearly much worse to come. Logically it also suggests that, with the best will in the world, achieving even a major reduction in man-made carbon emissions will not on its own cure the problem. All it could do would be to stabilise the concentration of CO_2 in the atmosphere at existing and unacceptably high levels. This would still be the case even if we were to aim for – and somehow achieve – the substantial 80 per cent reduction which many experts maintain should be the target.

In other words, and wholly contrary to the general impression conveyed by many news reports covering global warming and its effects, climate change is not like a light switch. It can't be turned off once the necessary political will has been found to encourage us to control and curtail our consumption habits. Nor, by extension, will the threat of it be extinguished even when we are forced to stop burning oil, coal and gas at their current rates because supplies of all three have finally run out.

Instead, if, as seems the case, the science is right, there is a near certainty that before long we will experience hugely destructive rises in sea level and some very extreme changes to our climate. This, even if governments around the world somehow agree on a meaningful limit for carbon emissions and succeed in pegging them at this punishing new level before sunrise tomorrow morning. Were that to happen – a wholly unlikely scenario – temperatures around the world would continue to rise for many years, and sea levels for many decades after that.

How long this time-lag might be no-one knows for certain. It could be in the region of fifty years or maybe as much as a hundred, but the reality is that it doesn't matter. Even without precise numbers we can say with certainty that the longer we delay taking some meaningful action – not to avert disaster but to safeguard whatever it is we most value, as individuals, families, communities or nation-states – the harder it will be to limit the extent of the damage and to secure our future.

Politicians too weak (and too scared) to act . . .
Put like that, things sound pessimistic in the extreme. But it is
not unrealistic, as can be seen when one asks: what is being
done now to dramatically reduce carbon emissions; and what
can we as individuals do to help? The answer to both seems to
be very little, and possibly nothing at all of any real, long term
significance. In 2007, for example, just $30 million of the US
Climate Change Science Program's $1.7 billion was spent on
evaluating how climate change will affect the population's
lifestyles and wellbeing.

Much of this lack of a meaningful response – as I write
the same USCCSP finds itself facing budget cuts – is due to
the lamentable tendency of politicians generally to avoid
proposing or taking any meaningful action, and also to their
time-honoured adroitness when it comes to following the line
of least resistance, particularly as now when they sense that the
electorate is likely to demand economic prosperity above all else.
Of course the public conspires to help them do this, with many
voters clearly keen to avoid hearing any bad news and many others
looking for solutions but prepared to accept only painless, cost-
free ones which will not damage or curtail their own lifestyles.

That said, when it comes to tackling climate change a
growing minority are apparently prepared to make changes to
their own patterns of consumption and behaviour, to which
end they are beginning to ask reasonable questions and
demand answers. Why do our cars still use so much fuel? Why
are energy-efficient cars not subsidised to make them more
affordable? Why are so many resources and materials
squandered on unnecessary packaging? Why isn't more of an
effort being made at government level to secure power from
cleaner, more sustainable sources? And, perhaps, most
tellingly, why is it so hard for even highly motivated
consumers to get the right sort of information they need in
order to make more socially and environmentally sound
consumer choices?

The answer is simple enough: governments are not doing their job properly. All too often, instead of showing leadership they pass the buck or take an easier if less effective route. (Just look at China and the US arguing over who is the most to blame.) Instead of taking courageous decisions likely to benefit everyone in the long term, politicians around the world cower behind the demands of short-term political expediency. And, crucially, instead of marshalling and assessing expert opinion honestly – and recruiting good scientific argument to their cause in the important battle for hearts and minds – they repeatedly cherry pick whichever data and research findings appear at the time to support their decision to do nothing of any real significance.

. . . and the experts don't help either

Unfortunately, in regard to this last aspect, the politicians are more often than not assisted by the recognised tendency of apparently independent experts to err on the side of the *status quo*, and to be over-cautious in their forecasts and findings. Frequently this is done because – as the highly regarded NASA physicist James Hansen reported in *New Scientist* – researchers and laboratories 'downplaying the dangers of climate change fare better when it comes to funding'. At the same time those who do not downplay bad news in this way frequently find themselves being pilloried for being alarmist and, as James Lovelock has observed, 'scientists are usually negative about new ideas'.

The result is that, instead of getting a range of different opinions about, say, climate change, one is left with a skewed distribution, and very little discussion or diversion either way from a largely conservative consensus. Elsewhere it has been observed that other professions act in a broadly similar fashion, including economists and statisticians, so that governments are never at a loss for something to fasten onto as an explanation – which is to say an excuse – for avoiding ever having to paint things as black as they really are, and therefore being forced to do something about it. (Sadly the public don't help matters either, choosing if only by their own inaction to be fed material by often biased parties instead of seeking out more reliable data and information themselves.)

For regrettable if understandable reasons politicians and the press also display a damaging tendency to prefer the simple over the complex. Crucial facts are often quietly disregarded or completely ignored. Sometimes, almost certainly, this is done to make an easy sell – the belief being that the public will simply not engage in a debate if it requires too much thought just to understand the issues involved. Similarly much of the relevant data will be ignored because of a belief that the basic concepts will be too hard to get across if every single factor is taken into account. (For this reason, for example, we hear very little about the extent to which the amount of heat which is naturally reflected away from the Earth will be affected now that the largest snowfields are shrinking; or the likely additional warming effect of the methane which we can now expect to be released as a consequence of melting permafrost in Alaska and the former Soviet Union.)

Everyone argues while the world burns

The consequences of this approach are easy to predict. On the one hand the few remaining climate-change sceptics can persuade the optimists to take comfort because the news is after all not so bad. On the other hand, more damagingly (as these simpler versions of the problems we face are more readily exposed to rigorous scientific criticism), the public is more than occasionally left with the sense that no-one in authority can be trusted to tell the truth, and that nothing they are being told is reliable or of any value.

At the same time the public can watch the two major polluters, China and the US, arguing about who is most responsible rather than seeking a real solution, both sides busily passing the buck rather than facing up to the consequences. And confronted by this, by a combination of government posturing and short-termism, a tendency to soft-pedal when it comes to bad news, by a lack of any genuinely meaningful action and now by an expanding array of new pseudo-green stealth taxes, many people simply switch off. Turning their back on the whole issue at the first opportunity, even in the face of catastrophe, many who could fight for change instead choose to do nothing.

Is it any wonder that so many of us simply roll over and go back to sleep? The arguments are indeed complex, the proposed solutions invariably inadequate. As an example, the EU's recent bid to reduce new-car emissions by a fairly substantial-sounding 20 per cent won't achieve much, not least because the new car industry is already lobbying hard to prevent any legislation taking effect before 2015 – and really the whole notion that we might actually travel less is so difficult to comprehend that most people simply choose to carry on as before.

Don't feel guilty – it's too late anyway

Maybe this is no bad thing. It's certainly not a good thing, but reality counts for a lot and the truth is that if we in the West don't use the energy someone somewhere else will. It is true too that, from an economic standpoint if not from a moral one, this might be even worse than if we use it ourselves. After all, the fact is that we in developed nations typically manage to squeeze far more value from a given unit of energy than those in the developing world. Not just because we have more efficient machines, but also because our cars, planes and trucks tend almost invariably to pollute less per mile of travel than the older or less sophisticated ones employed in poorer parts of the world.

We can argue about that, of course, particularly when it comes to the questionable allocation of this energy – for example, producing large four-wheel-drive vehicles which are then driven around western cities rather than using that same energy to raise overall living standards in Asia or sub-Saharan Africa by boosting industry and employment. But against this is the realisation that it is now simply too late to stop all the bad things happening. Too late to prevent global warming, and too late to head off the change in our climate that this will necessarily lead to. We are, in this sense, no longer the masters of Mother Nature but just another aspect of it.

Unsurprisingly that last bit is something few people want to talk about. The green lobby is reluctant to put it in these terms for fear of sounding defeatist or apocalyptic at the precise moment they find themselves, finally, manoeuvring into the mainstream. And politicians certainly won't put it like that either, because then they would have to do something concrete in order to deal with the threatened changes. Even the IPCC prefers to choose its words carefully, although it's now more than two years since chairman Rajendra Pachauri warned that by doing little or nothing 'we are risking the ability of the human race to survive', and more recently an English delegate on the panel admitted that the EU's bid to limit average temperature rises to 2°C is almost certainly not going to succeed.

Instead the reality is that we now need to shift the focus from prevention to protection, and to be proactive instead of merely reactive. It is, in short, no longer simply a matter of arguing about how and when to implement changes such as those demanded at Kyoto, or of fitting solar panels to our homes or deciding to drive smaller cars on fewer journeys. Rather we need to move into a new and more energetic response phase. Planning ahead to cope with new and different climate patterns and for a life in the post-carbon era by strengthening emergency and support service capabilities, evaluating how we can most effectively deal with change and relocate literally thousands of newly vulnerable communities, and – as a part of this – to prepare our own towns and cities for a massive influx of new environmental refugees.

Chapter 3
THE FUTILITY OF GESTURE

With a problem of this magnitude, every little doesn't *help.*

Of course even given the most gloomy prognosis there are those who are genuinely willing to make changes to reduce their own environmental impact, and others who are already doing so. But for these people, unfortunately, the situation is far from easy. For one thing, the necessary calculations are not easily made; for proof of this just look at some of the laughably inadequate 'carbon calculators' popping up on the world wide web. For another, many individual responses, while sincerely made and looking at first glance quite sound, in the end prove if not quite damaging then certainly worthless when set against the enormity of what is, after all, a truly global problem. Not just because, with a problem of this magnitude, every little *doesn't* help, but also because the unintended consequence of unguided actions can be to create even more problems than one had before.

For example, and as noted previously, switching the light off when you leave a room is not going to achieve anything worthwhile, and certainly not when the average British family has an estimated ten to twelve additional electrical and electronic gadgets around the home, a majority of which are kept on 'standby' rather than being switched off. According to the Energy Saving Trust, this habit alone wastes the equivalent amount of energy to that produced by two and a half 700 megawatt power stations each year. Indeed, according to the organisation's chief executive Philip Selwood, by 2010 an amazing 45 per cent of all domestic energy in the UK will be expended on such devices, a measure equivalent to the entire output of fourteen power stations.

Greater wealth means more pollution

This situation is being made even worse by the growing number of single-occupant homes, and the same basic pattern can be seen throughout the domestic arena. It is true, for example, that some measurable gains in efficiency are being made across a range of different domestic appliances – dishwashers which can operate at lower temperatures, washing machines which require less water than conventional models – but any theoretical gains made in this way are more than offset by the fact that more of us have more of them than ever before.

Similarly sales in Britain of patio heaters are currently racing ahead, this despite the fact that a typical model emits as much CO_2 as one and a half family cars. At the same time

Formula One enthusiasts have been enjoying the somewhat bizarre spectacle of the Honda team launching its own environmental initiative and ditching its sponsors' logos in favour of a big picture of Planet Earth. 'We're not saying F1 is green,' said a spokesman, in a statement which came as no surprise to anyone, 'but if 1 per cent of the people who watch F1 were to change a light bulb for an energy-saving one it would save 38,000 tonnes of CO_2.'

Of course that's not much given the current global total of carbon emissions of around 7 billion tonnes annually, and actually it's not much compared with Formula One's own hefty carbon footprint either. More significantly perhaps, especially given continuing government encouragement for us all to make the switch to these allegedly more efficient bulbs, the whole low-energy lights thing is itself a complete red herring. Most obviously, saving a few watts per room is simply too marginal a change in overall energy usage to make a significant difference – and that would be true even if we were all to make the change. In addition, in making the argument for the switch away from incandescent bulbs, no-one has allowed in their calculations for all the extra energy, engineering and materials required to manufacture the far more sophisticated low-energy replacements.

Legislators love a gesture . . .

In fact when it comes to this single domestic product there are so many other considerations which have to be taken into account that their sheer number provides an excellent illustration of the incredible complexity of climate change; also of how and why it is that this complexity eventually conspires to ensure that the vast majority of well-intentioned gestures will almost certainly fail to achieve anything meaningful.

For example, because 80 per cent of the power consumed by a conventional incandescent bulb emerges as heat rather than light, a switch to cooler-operating, lower-energy light bulbs throughout a building will simply be counteracted by the central heating system having to work harder. (Your home's heating is thermostatically controlled, so without any deliberate human intervention the system will automatically compensate for the heat-loss once these new bulbs are installed.) Of course householders and office managers could put on an extra sweater, and avoid this unwelcome consequence by turning down the thermostat when the new lightbulbs are installed. In the unlikely event of their thinking to do this, however, it will merely dim the lights further – as low-energy bulbs are markedly less efficient at lower temperatures, hence the recommendation not to use them outdoors.

Finally, of course, if at the same time a homeowner were to switch to one of the new 'green' utility suppliers any marginal CO_2 savings promised by the low-energy lightbulb would disappear altogether. Because the new generation of utilities producing power from wind, solar, wave, geothermal, hydro-electric or nuclear sources – precisely the sort of sustainable alternatives likely to be favoured by the same, environmentally minded citizens who switch to low-energy lightbulbs – claim to emit very little or no CO_2 anyway, there would simply be no savings of the gas to be made.

. . . but 'low energy' lighting is just a distraction

Not, mind you, that any of this has prevented governments around the world from continuing to promote low-energy light bulbs as a real solution; indeed it seems certain that EU member states will soon follow Australia and move to outlaw conventional ones entirely, thereby forcing us all to switch regardless of the minor benefits, or more probably negative effects, that such a switch will bring. Because of this it is hard to avoid the conclusion that the impetus behind such a move really has more to do with ticking the right 'carbon-reduction' box than with reacting rationally to achieve any meaningful change, or with doing something because it is easy rather than because it is effective.

After all, banning conventional light bulbs in this way is action of a sort and as such one which, politically speaking, is sure to be much more acceptable than attempting to price people out of their cars or off aeroplanes, although either one of these moves would clearly make a far larger dent in any given country's overall carbon emissions. Which brings us to the car. As with the low-energy bulb, buying a new, more economical model is an easy concept to grasp and one which appears to make sense – or rather it might if enough of us were encouraged to downsize by the use of genuine incentives in place of the present government's so far somewhat feeble efforts. You know the kind of thing: wholly token efforts such as raising the road fund licence by a few tens of pounds for anyone considering spending sixty or seventy *thousand* pounds on a lavishly equipped, over-engined gas-guzzler.

Cleaner cars don't help either

Once again, though, the truth is far harder to unravel. For one thing the contribution to global warming made by British motorists is in real terms very small indeed. Small compared to the emissions from our homes, and very small compared to those coming from the construction and manufacturing industries. It is also very, very small compared to those which are soon to come from literally hundreds of millions of Chinese and Indian motorists who, in rapidly growing numbers, are beginning to drive private cars of their own, the vast majority of which can be expected to fall far below the standards of anything which could be legally built, sold or driven in North America or here in the EU.

In fact, according to Professor Julia King, the Midlands academic recruited to head Gordon Brown's 2007 low-carbon car review, road transport emissions in total account for just one fifth of all UK CO_2 emissions. Moreover, much of the worst of this transport-related air pollution comes not from privately owned cars at all but from the country's ageing bus and commercial vehicle fleets or as a consequence of worsening traffic congestion and increasing mileages, rather than actual poor vehicle engineering.

There is also the argument that Britain is in any case only one small country among scores of other more substantial polluters – although of course it can and is being argued here that as a relatively rich nation we are in a position to set a good example for the developing world to follow, and should do so. That said, there is of course no indication or reason for thinking that this is a lead which anyone would follow. At the same time it is only a gesture, and if the previous chapters do nothing else they should at least underline the fact that the time for gestures is gone and that we need to take some real action.

'New for old' simply adds to the problem

Supposing we leave all this aside, however, and go and buy that smart, new, more compact, more fuel efficient car anyway. What then? Our old car will obviously need to be scrapped, the logic here being that it is so unacceptably thirsty and polluting that the only rational thing is to get rid of it altogether. That's quite a thought, however, when one considers that, in Britain alone, a prosperous country, more than fifteen million cars, nearly one in every two, are seven years old or more. Of these, according to published Society of Motor Manufacturers and Traders (SMMT) figures, nearly nine million have passed their ninth birthday and just over four million are more than twelve years old.

Even assuming these vehicles are well maintained, goes the scrap-'em argument, the majority will not have the latest emissions control equipment and will also be considerably less fuel efficient than their modern equivalents. And this is true: a 1995 Vauxhall Astra 1.7D, for example, averages 44.2mpg,

whereas its replacement (the 1.3 CDTi) returns 58.9mpg.
Since every litre of diesel saved in turn saves 2.6 kg of
atmospheric CO_2 a driver covering 12,000 miles a year in the
newer car could in theory save himself in excess of £300 a year
in fuel costs and enjoy the feel-good factor of preventing a
whopping 800 kg of CO_2 going into the atmosphere each year.

But hold on a minute. Besides this being just
another example of an almost pointless local reduction of
demand – and of course it is important that any money
thus saved is not spent on something else which directly or
indirectly produces CO_2 – there is surely something
fundamentally wrong with seeking to destroy a perfectly
good, fully functioning machine. Apart from anything else,
before scrapping anything in this way, one needs to take
into account the very substantial environmental cost not
just of using it but also of creating it in the first place, and
indeed of getting rid of it afterwards.

The damage is done before you even drive

Every year, after all, quite phenomenal quantities of many
different resources – not just iron and steel but also increasing
amounts of very expensively refined aluminium, together with
glass, water and countless chemical solvents – go into building
anything up to fifty million vehicles around the world. To this
must be added the substantial volume of oil required to make
the increasing proportion of plastic components in modern
cars, and yet more of it – with coal and gas too, depending on
where the cars are made – to fuel the many hundreds of car,
bus and truck factories around the globe and by extension
those of their many thousands of suppliers. Also, of course, the
immense distribution requirements of such a widely dispersed
industry, which daily consumes yet more reserves in order to
fuel fleets of transporter ships to bring the new cars to market
and the trucks needed to deliver these cars from the docks to
the dealers.

It is little wonder that the figures computed by the Environment and Forecasting Institute in Heidelberg make such grim reading. They show that long before an average, medium-sized family car even reaches the showroom it is already responsible for significant amounts of damage to air, water and land ecosystems through the extraction of the raw materials needed for its manufacture. Thereafter the same average car will account for some 26.5 tonnes of waste materials and some 922 cubic metres of polluted air, the cradle-to-grave calculations also allowing for a mammoth 102 million cubic metres of polluted air arising as a consequence of the car's eventual disposal at the end of its useful life.

In Britain, speaking for the car makers themselves, the Society of Motor Manufacturers and Traders admits to some equally staggering figures, and this despite the UK quite clearly having some of the cleanest and most efficient car factories in the world. They cite 0.6 tonnes of CO_2 per car built together with 2.3 megawatt/hours of energy and 3.2 cubic metres of clean water – all this without the car even turning a wheel, meaning one surely has to question the wisdom of replacing existing perfectly usable cars with brand new ones sometimes many, many years before the former have actually worn out.

Industry improves – but nowhere near enough

Not unnaturally, in response to all this, car builders speaking through organisations such as the SMMT are keen to highlight the fact that their environmental record is improving and that a growing proportion of every new car is now recyclable, at least 75 per cent of it by weight but as much as 90 per cent depending on the model – although the reality is that in many cases actual markets for all this recycled material have yet to be identified. The same industry likewise cites an average distance per car across the EU of only 15,000 kilometres annually, and the fact that CO_2 emissions from new cars have fallen from an average of 185g/km in 1999 to just 167g/km seven years later.

Such figures, of course, suggest an improvement on the situation which pertained a few years ago, but the reality is that such has been the rise in the number of new cars over the same period that the total distance travelled and the resulting emissions – the only figures which actually matter – are regrettably much, much higher. Once again, any apparent efficiency gains have rapidly been nullified by the fact that so many more of us have cars than was previously the case. Indeed, at the time of writing, there were more than thirty million cars on the roads of Britain alone – compared to fewer than nineteen million in 1983 – and a stretch of the M6 was being widened at a cost of £3 billion, or approximately forty times the sum being spent by the government on its much vaunted low-carbon buildings programme.

The Prius problem

Nor, contrary to popular myth, are petrol-electric hybrids such as the million-selling Toyota Prius going to provide a viable answer either, although this may come as little surprise to anyone who finds suspicious any claims that a car with two motors could be greener and more efficient than a car with only one.

In their defence it is true that by generating their own electricity and running on battery power at slower speeds around town the various hybrid models being championed by Honda, Toyota and Lexus successfully reduce emissions in the urban environment. All of them are excellent cars too, particularly the largest Lexus, and their green credentials are more credible than those of plug-in electric cars – which merely transfer the pollution to the power station chimney.

But against this, the fully hybrid future which some posit as a likely solution would require much greater reserves than we actually have of the various key minerals needed to ramp up storage battery production. Supporters of hybrids also choose to ignore the cost of the embodied energy in such vehicles, i.e. the energy cost of actually building them – which in the case of the Honda Accord hybrid has been put at more than 50 per cent higher than for the conventional model. A hybrid future would also require us to find a way to recycle more efficiently the waste materials generated when the batteries eventually expire, an aspect of the otherwise highly popular Toyota Prius which generally goes unremarked.

Even if we do all this, however, and whatever we choose to do with our older, dirtier vehicles, we are still left with the obvious insanity of routinely employing approximately 1 tonne of steel, plastics, glass and resource-hungry electronics in order to transport a single human being from A to B five days a week.

All roads lead to ruin

Transportation, at all levels and in every industrialised country, poses one of the biggest and most intractable problems of our age. Given that roughly a third of global energy output currently goes toward moving goods and humans around, simply 'doing something' isn't going to make a big enough difference. In truth the only sensible answer is not to travel at all.

Put like that it sounds pretty draconian – but only until you sit down and think about the end game. It is then that one realises the full absurdity of much of what we are being asked to swallow these days. For example, a company such as BP showcases the good news that it has slightly reduced its carbon output by changing the paint it uses on its tankers, while apparently ignoring the vastly bigger emissions produced by the contents of those very same tankers. Or *The Times* carries a long feature telling readers about 'seven simple ways you can make an immediate and positive impact on our environment' but runs it alongside another called 'The Art of Travel', which encourages those same readers to 'discover a variety of enjoyable road journeys across the UK and win a luxurious family long weekend in Edinburgh with Renault'.

Train versus Plane – well, who cares?

The truth is they and we are just tinkering around the edges. Consider the biggest debate of them all, the one about trains versus aeroplanes. We are all familiar with airline spokesmen insisting that their aircraft fleets are getting cleaner and greener, while refusing to acknowledge that because there are so many more aircraft up there the overall effect is much, much more pollution than the year before. Boeing, for example, claims its new 787 'Dreamliner' is 20 per cent cleaner than a conventional jet, but even if this is correct and if everybody switched to the Dreamliner the fact that aviation is growing at 5 per cent per annum means that any theoretical gains will be completely wiped out within just four years.

Supporters of railway travel similarly spend their time demonstrating how much greener trains are than planes – at least in terms of their best-case CO_2 emissions per passenger mile. Indeed, this angle is so much taken for granted that even aviation industry spokesmen don't bother arguing with its basic premise. Nevertheless, anyone making the claim willfully ignores the fact that the figures only stack up if every train is a full train – something which, given the staggering expense of train travel when compared to air travel, and the impressive record the aviation industry has when it comes to filling aircraft, is rarely the case outside the rush-hour. Efficient trains are also fiendishly expensive. The annual French subsidy, some 2 billion Euros, would be sufficient to buy every rail passenger in France a new car every year.

Finally, for their part, environmental protesters continue to demonise the building of every new runway, yet refuse to consider the similarly massive environmental impact of building and maintaining hundreds of railway stations, of constructing scores of bridges and tunnels, of laying and relaying many thousands of miles of railway tracks and overhead lines, and of manufacturing and maintaining literally millions of concrete sleepers up and down the country.

The answer? Don't travel!

In truth, though, it's a stale debate and a pointless one, because however you choose to travel – and more importantly however the rest of the world chooses to travel – if the emissions issue does not get you the availability of fuel (or lack thereof) will, and soon.

Chapter 4

CARBON CAPTURE AND SEQUESTRATION

*It is always going to cost money to capture something. It isn't
a free exercise.*

The theory behind carbon capture and sequestration
is straightforward enough: that the main pollutant is
not fossil fuel itself but rather the atmospheric CO_2
which collects every time we burn that fuel.
According to would-be sequestrators it therefore follows that,
assuming one can find an efficient way to capture this
atmospheric CO_2, and to safely contain it somewhere, we will

be able to avoid or perhaps more accurately mitigate some of the worst effects of global warming.

Of course it's not quite that simple. It is true that CO_2 can be captured and removed from the atmosphere. It is true as well that the gas can be broken down and rendered harmless by the action of plants and trees, which by the process of photosynthesis automatically convert it into carbohydrates for growth and oxygen. It is also true that the gas itself can be and is already being absorbed in great amounts by the oceans, chief among the planet's most effective and expansive natural carbon sinks.

But each of these stages poses its own major challenges, and there is absolutely no certainty that these can be overcome at all, let alone within the necessarily very short timescale that is allowed us by the present crisis. There is also the almost comically perverse aside that while, on the one hand, the hunt is on for a viable way of storing carbon in the ground, on the other we are busy looking for ways to avoid doing precisely this.

Take the issue of plastic packaging and plastic bags, for example, which particularly in Europe have found themselves to be very much Public Enemy No. 1 in the great debate over the horrors of landfill. Because of this, for the last twenty years or so society has been seeking ways to exchange conventional plastic for biodegradeable alternatives, which is to say formulations that function as plastics when first made but decompose more quickly and more completely once they have been discarded and put into the ground. Unfortunately when such materials decompose they do so into CO_2 and methane – that is, the two most high-profile greenhouse gases – whereas conventional supermarket packaging and carrier bags remain underground as complex plastics for many hundreds of years, thereby releasing very little CO_2.

In other words they function as small but efficient carbon sinks. Nevertheless we are working hard to destroy them and to replace them with something which will in turn generate even more greenhouse gas. Greenhouse gases that we will then have to capture and store somewhere else.

First catch your carbon

Whether at the power station chimney or at the factory gate, carbon capture is something which is far easier said and understood than actually done. On the positive side the problem is by definition reasonably localised and well understood from a technical standpoint, and much of the technology required is already available in some form. On the other hand, however, and as Harry Audus, an engineer with the International Energy Agency's greenhouse gas programme, pointed out as long ago as 2005, 'it is always going to cost money to capture something. It isn't a free exercise.' In fact the costs of carbon capture on a meaningful scale are likely to be astronomical, in both financial terms and in terms of unavoidable energy loss.

Part of the problem is that CO_2 is just one of many different waste gases escaping a power station chimney. A modern power station burning natural gas or methane (CH_4) also produces hydrogen, so before the CO_2 can be used or stored it needs to be separated from the hydrogen and from any other contaminants which will invariably be present. Fortunately there are means of doing this, and so-called amine scrubbers which chemically isolate the gas have already been fitted to a small number of power stations in the US. This has been done in the hope that the CO_2, once isolated and compressed, can be sold off as a pure chemical product in order to partly offset the cost of this expensive process.

The majority of power stations, however, particularly the older ones and those being built in developing countries (notably China) are not only much less efficient and far more polluting than these modern, gas-powered units but also far harder to convert to use this sort of scrubbing technology. Additionally, should the conversion be made anyway, their efficiency will drop still further, as the various pumps and so forth required by the amine scrubbers and the compressors are themselves very considerable users of power. So much so, indeed, that the cost of isolating the gas, compressing it, cooling it, transporting it to the sequestration site and depositing it there will actually consume virtually all of the energy which was generated by the power stations in the first place.

In a bid to counter this, two newer processes have come under the spotlight more recently, processes which chemically clean the fuel before it is burned in power stations. Through a mechanism known as gasification, for example, coal can be turned into hydrogen gas; alternatively, the new generation of 'oxyfuel' plants effectively break down ordinary air so they can use just the oxygen component to burn the coal. While new and relatively untested, both are said to produce a more readily captured form of CO_2.

Yet even by generating power in this way one is left with the not inconsiderable cost of compressing, cooling, transporting and depositing not just millions but billions of tonnes of gas every year once it has been isolated. The carbon-capture concept also fails to address the approximately 10 per cent leakage lost to even the most efficient plants, and the fact that a huge proportion of the most harmful CO_2 emissions – thought to be around 60 per cent of the total – comes not from industrial chimneys at all but from other sources, such as offices and private homes where capture is much harder, and jet aircraft and other vehicle exhausts.

Store it – but where?

Assuming these aspects can eventually be overcome, storage is in theory a less demanding problem technically. That said, and because the volumes involved are so immense, it is clearly not possible to simply build our own storage facilities as we have done for low-level radioactive waste. Instead there is a requirement to find vast natural storage facilities. Among those currently being considered and evaluated by organisations such as Edinburgh University's School of Geosciences are scores of old oil and gas fields. Those in the North Sea, for example, are otherwise nearing the end of their useful lives.

Such a solution sounds attractive – neat and nicely self-contained. There are nevertheless a number of problems regarding the long-term viability of these natural carbon stores, and in the short term at least some legal problems to consider as well, many of which are discussed on the university's dedicated website, www.co2capture.org.uk.

To begin with there are presently any number of European treaties in force which almost certainly make it illegal to dump millions of tonnes of what is after all a type of industrial waste beneath the North Sea. Admittedly for several years now the Norwegian Statoil company has been pumping many millions of tonnes of its own CO_2 emissions into a sandstone layer below its North Sea territories, but this is allowed because while it is in effect being stored, the waste gas is primarily being employed as a propellant to squeeze oil out of the ground in the opposite direction. Aside from this issue of definition, however, so far at least there have been no problems with leakage, although geologists are still on hand to monitor this possibility.

A similar programme is also underway in Scotland, with CO_2 which has been isolated from the emissions of a power station at Peterhead being sent offshore along an existing pipeline to the Miller oilfield. In common with other fields in the North Sea this one is nearing the end of its life as a viable oil-producing area, even though in this particular case around a third of the reserves still remain untapped. It is hoped that once the scheme is fully up and running, by 2009, the CO_2 which is isolated using a special membrane may be used to make the extraction of this last, costlier third easier and more cost-effective. Once this has been verified, the gas can then be stored in the oilfield itself, some 4 kilometres below sea level.

Unfortunately, however, this is only ever going to be a partial solution: doing something similar on the much larger scale that would be necessary to make a difference would also be prohibitively expensive. So expensive, that is, that it goes a long way towards explaining the marked reluctance on the part of the oil companies to get involved, even though it is they who own, operate and control the necessary infrastructure of rigs and pipelines which would be needed to make it happen.

There is, it is true, a second type of geological store, within the porous rocks known as saline aquifers which lie deep below the surface in sedimentary basins. Full of salty water (which as such is good for neither drinking nor agriculture), these have survived intact for millions of years and are now estimated to have sufficient capacity beneath the UK alone to store more than 7 billion tonnes of CO_2 or about 500 years' worth of the UK's emissions at current levels. Here again, though, there are hideous cost implications and some very considerable technical difficulties, in that such storage sites are far less well understood geologically speaking than the aforementioned gas and oil fields.

Nevertheless, say supporters of such schemes, if the political will were there the cost could be offset to a great degree simply by charging everybody more for the fuel they use. However, with even the best estimates putting the cost of doing this at an enormous (and potentially politically destabilising) 2 per cent of global domestic product it is difficult to see from where the impetus will come to put such a scheme into operation.

The sea – our saviour?

Because it is known that something between a third and a half of the CO_2 in the atmosphere automatically dissolves into the world's oceans, consideration has also been given to discovering a more affordable means of increasing the percentage stored in this way.

This natural uptake of CO_2 depends upon two different but connected mechanisms known respectively as the physical and biological pumps. In the first of these CO_2 automatically dissolves into the oceans at higher latitudes before being carried down by sinking currents into the ocean depths, where it is thought to remain for hundreds of years. The effect of the biological pump, meanwhile, is in part due to phytoplankton activity or more specifically photosynthesis, which converts the gas into harmless organic matter and oxygen. Animals then eat this organic matter (micro algae) and when they in turn die a proportion of their body mass sinks to the sea floor, where together with its embedded CO_2 it is eventually buried beneath sediments, one hopes for many millions of years.

To take advantage of this potential carbon sink and the natural functioning of the physical pump it has been proposed that CO_2 be bubbled through sea water on a mammoth scale in order to increase the uptake of the gas. Another option might be to capture and liquefy the CO_2 before transporting it by pipeline or ship to somewhere where it could be safely injected at great depth, where it would remain isolated from the atmosphere for many centuries.

Unfortunately, however, there are signs that the world's oceans are already failing to keep pace with the growth in atmospheric CO_2, as already mentioned. The capacity of one of the most effective, the Southern Ocean around Antarctica, is thought not to have increased at all since the 1980s – despite the fact that CO_2 emissions have themselves risen by around 40 per cent over the same period. One also needs to consider that were additional gas to be dissolved in the oceans the resulting cocktail would not actually be inert. CO_2 added to seawater produces carbonic acid, which as the name suggests would badly upset the pH balance of the oceans and be potentially disastrous for overall marine ecology.

In truth the effects of this are not yet precisely understood, but they can certainly be guessed at particularly if – as would be necessary to score any lasting benefit – such processes were carried out on an appropriately global scale. In an attempt to balance this it has further been suggested that any increased acidity could in turn be neutralised by dumping quantities of dissolved carbonate minerals into the sea. In theory this should allow even more CO_2 to be absorbed from the atmosphere – or rather it might were it not the case that sufficient sources of limestone and other suitably soluble alkaline compounds have not yet been identified to match the scale of current and future CO_2 emissions.

Seeding the oceans

Another alternative to simply dumping CO_2 into the oceans might be to look instead at the biological pump and to seed the waters with additional minerals, in particular iron and urea or nitrogen. This would make it even more efficient than at present, and also serve to offset the fact that global warming is presently contributing to the loss of a full 1 per cent of all algae and other marine plants annually, flora which under normal conditions would actually be working to absorb carbon.

It has been noted, for example, that although the aforementioned Southern Ocean is rich in nutrients it is home to relatively little phytoplankton. This is thought largely to be because the seawater is deficient in dissolved iron, an element which in this particular region is essential for large-scale phytoplankton growth. Because of this a number of investigators have proposed fertilising the waters with more iron in order to nourish larger populations of the microscopic algae.

More calcium carbonate-secreting phytoplankton, researchers say, would allow for greater absorption of CO_2 through photosynthesis, and ultimately for many millions of tons of CO_2 to be carried down into the ocean depths. Assuming this material remained there for hundreds or even thousands of years it would no longer contribute to global warming. That's quite a thought, particularly if the claim of one company active in the area is true. Speaking to the *Sunday Times* in late 2007 Planktos, a US-based corporation, calculated that one ton of iron dumped into the sea could remove 100,000 tons of CO_2.

The numbers are impressive and although the science behind the theory is by no means proven there have been a number of studies by Planktos and other organisations. Stéphane Blain and a team of oceanographers from the University of the Mediterranean in Marseilles, for example, monitored the growth of a phytoplankton bloom close to the Kerguelen Islands, fed by a natural welling-up of iron and other nutrients from deeper waters. Lying roughly equidistant from Africa, Antarctica and Australia, the bloom was truly immense and over a period of three months grew to cover more than 17,000 square miles. There are obviously less desirable impacts on the ecosystem of such a development, but Blain and his team were nevertheless able to demonstrate that the amount of carbon taken out of circulation (as the phytoplankton dies and sinks to the ocean floor) was surprisingly large: anything up to 100 times more per unit of iron than had previously been estimated.

Nor is the Southern Ocean unique in this respect: in fact great swathes of the world's oceans, possibly as much as one-third of the total surface area, have little or no biological activity at present because the growth of the relevant phytoplankton is limited by a similar lack of iron. Indeed, a number of relatively small experiments in recent years – and in various different locations – have managed to demonstrate that seeding them with iron works and can result in a marked increase in photosynthesis.

That said, caution needs to be exercised with regard to where such programmes are carried out. Not just because there have so far been no authentic peer-reviewed studies to authenticate the process, but also because such schemes are highly likely to affect other marine activity in the area and thus, for example, reduce commercial fishing yields. Additionally the formation of calcium carbonate in this way can actually release CO_2 into seawater: as calcium carbonate is soluble in seawater, the CO_2 is eventually freed, to resurface as a consequence of normal oceanic upwelling. It seems probable too that only the formation of sediments containing plankton-derived organic matter would be able to remove CO_2 in sufficient quantities, and it is certainly the case that only a tiny proportion of the phytoplankton bloom – perhaps as little as 5 per cent – actually sinks to the very bottom of the ocean.

Whatever the effectiveness, however, operating such a carbon mitigation strategy on a global scale would require the volumes of iron to be considerable; also the cost – prompting the question who pays, and indeed who gives permission? – although the good news is that the same programme could theoretically pay for itself by eventually generating an economic benefit as well. An increase in the sea's overall productivity could in time lead to a harvest of both food and energy for the operators of the scheme.

In turn this means that a series of international treaties and agreements would be required in order to ascertain who pays for the set-up of the scheme, and eventually how the spoils are divided once the current international waters have been redesignated as economic exploitation zones. That's assuming, of course, that everyone is happy with the plan to change an entire ecosystem in this way and on such a scale. . . .

Might the future grow on trees?

It is well known that, over the long term, trees and indeed the soil itself are efficient absorbers of CO_2 from the atmosphere – so would it help if we planted more of them?

In fact it might be more sensible to ask whether or not it would help if we stopped cutting them down in such numbers, since on present trends the effects of the deforestation which is expected to take place between now and 2010 is thought to be roughly equivalent to the total volume of aircraft emissions from the Wright Brothers' first flight through to 2025. One also needs to consider that trees don't simply absorb CO_2 as was thought for a long time, but actually emit more carbon over time than they actually absorb.

Leaving that aside, the reality is – even without the additional squeeze being put on productive land by the inexorable rise of biofuels (see Chapter 8) – we simply don't have enough land on which to plant the numbers of trees we would need to make a useful difference. Dealing with the normal emissions from an average European, for example, at approximately 10 tonnes of CO_2 per year, would reasonably require 5,500 square metres of managed forest – equivalent to almost 1.5 acres per person – together with somewhere to store the dry wood harvested each year from that forest.

That's a lot of forest and an awful lot of dry wood. To put it into perspective, the United Kingdom has only 4,000 square metres of land per person – or around one acre, including all the land currently occupied by housing, roads, industry and so forth – which is clearly not enough. And taking the planet as a whole such a scheme would require almost 10 per cent of all agricultural land to be made over to managed forest, and for a whopping 1.4 billion cubic metres of dry wood to be stored annually.

All the wood has to be stored or used to make something reasonably enduring, because if it is burned as fuel or even allowed to rot into the forest floor the carbon contained within it will simply be released back into the atmosphere. For this reason it is also important to have the right kind of forest. New growth, meaning a young forest comprised of saplings, will under normal circumstances sequester relatively large volumes of carbon proportional to its total biomass, but actually stores relatively little because each tree is very small. By contrast, a mature or old-growth forest functions as an efficient reservoir of carbon, holding large volumes, although its capacity for absorbing more on an ongoing basis will not be that great because older trees grow more slowly.

For these reasons any forest being used to sequester and store CO_2 needs to be managed full time. Natural, unmanaged forests, such as those of the Amazon basin, certainly perform a useful function, but they also ebb and flow in response to a number of factors that form part of the natural ecological system. Disease, wind, fire and other wholly natural occurrences can cause trees to die individually or *en masse*, an event which automatically results in the release of CO_2 back into the atmosphere, even if it is then followed by a process of regrowth – thus beginning a new process of carbon build-up within that same forest.

There are other issues too. Scientists now think, for example, that not all forests are efficient at trapping CO_2 especially if the trees are growing at higher latitudes. New research indicates that any benefits accrued as trees reduce atmospheric CO_2 may be outweighed by those same trees' capacity to trap heat closer to the ground. In fact computer models suggest that trees really only work to cool the planet if they are planted in tropical latitudes, although even here if temperatures continue to rise the process can stop altogether once the leaves close their stomata (the CO_2-inhaling pores) in order to avoid dehydration. It is also now thought that warmer weather, reduced rainfall and a greater number of cloudy days – all recognised features of climate change, of course – are serving to slow tree growth in many key forested areas.

All in all, then, the situation is unclear, indicating that at very best sequestration of this sort can never really be anything more than a minor or temporary mitigating activity.

The car rides to the rescue?

Should we be surprised that the hunt is on to find a new use for all this CO_2 instead of just dumping it into the sea or into managed forests?

One such is the somewhat unexpected alliance between a number of leading auto industry suppliers and Greenpeace, a group which has been lobbying the world's car makers in the hope that they will consider using this waste gas as a refrigerant in their cars' air conditioning systems.

According to an organisation called Alliance for CO_2 Solutions, doing this has 'the potential to knock out 1 per cent of global greenhouse gases'. Indeed CO_2 technology, says Wolfgang Lohbeck, Greenpeace's Head of Technical Innovations and Projects, 'outperforms competing chemical refrigerants on all three counts. It is more environmentally friendly, more technically ready, and more cost-efficient than competing chemical refrigerants.'

Little wonder then, that with the market for such refrigerants worth an estimated $14.5 billion a year, the race is now on to find an alternative to the existing refrigerant HFC-134a. (This is said to have a global warming potential more than 1,400 times higher than that of the new CO_2-based technology, and an EU directive requires it to be phased out by 2011.)

Any move within the motor industry is especially significant because many other industries are reportedly adopting a 'wait and see' approach, the suggestion being that, if the motor industry adopts CO_2 as the new standard for its systems, manufacturers of vending machines, supermarket chill cabinets, heat pumps for domestic water heating, industrial refrigeration and so on will likely follow suit.

Were that to happen the saving could, say analysts, rise to 3 per cent of the global CO_2 problem. That is hardly sufficient to avert disaster, but there is certainly something attractive about taking a problem and turning it into part of the solution – particularly when the other chemical alternatives would simply add to the problem, as they have to be manufactured from scratch with all the accompanying environmental burdens that this suggests.

Chapter 5
THE CARBON OFFSET CON

*It's like announcing you are going on a diet and then paying
someone else to eat less on your behalf.*

The continuing confusion over the whole issue of capture and sequestration is certainly something to think about when considering the growing interest around the world in carbon offsetting and carbon trading – or as Channel Four's *Despatches* programme preferred to bracket both types of scheme in a July 2007 broadcast, 'the Great Green Smokescreen'.

The theory behind carbon offsetting is simple enough, which presumably goes a long way towards explaining its current popularity. It is also relatively cheap compared with the complexities of carbon capture, extremely quick and easy to explain to a lay audience, and commercially easy to administer once a scheme is up and running. As a result of this happy combination, and in what has become a fashionable initiative in many different countries and across several key industries, consumers worldwide are now being encouraged to offset their harmful CO_2 emissions rather than reducing them.

The key to doing this, they are told, is creating or contributing to an ecologically beneficial scheme somewhere else as a way of compensating for large additions to one's own personal carbon footprint. At its simplest this means that if your family car or continental holiday emits, say, 1 tonne of CO_2 in a given period the harmful effects of this can be theoretically offset by – for example – planting sufficient trees to absorb approximately the same amount of CO_2 per year, or by contributing to an ocean-seeding scheme of the sort described in the previous chapter.

A good example of this is the way in which delegates flying to Scotland for the 2005 G8 summit at Gleneagles were presented with certificates assuring them that the emissions from their flights to Edinburgh and back again had been entirely mitigated by a programme that paid for the tin roofs of shanty-town huts in Cape Town to be replaced with a better, more insulating material.

Big corporations and even entire countries can join carbon offset schemes too, or for that matter create their own. Honda, Volvo, Avis and Dell are just some of the household names now associated with major tree-planting schemes, while the HSBC bank claims that offsetting means it is already 'carbon-neutral', Marks & Spencer says it will be by 2012, and recently have come suggestions that the whole of Britain could be 'carbon neutral' by 2050. On a more modest and seemingly more realistic scale a number of Mexican pig farms – which capture methane emissions from the animals' waste instead of letting them leak back into the atmosphere – receive funding from the oil giant BP as part of its offset programme.

Guilt becomes a business opportunity

As a consequence of this and other schemes the World Bank calculated that the value of such schemes doubled globally in 2006 to reach a staggering $5 billion. Given this, it is perhaps unsurprising that – having identified a source of free money apparently growing on trees – growing numbers of specialist companies are now springing up to facilitate these various offsets and so ameliorate public feelings of guilt. With our assistance, say the companies, consumers simply write out a cheque instead of having to buy the land and plant the trees themselves – although the reality is that none of these schemes is as straightforward as it seems, or anywhere near as effective as supporters and operators claim.

BP's website, for example, at first suggested that the aforementioned pig farms could offset each year the effects of 750,000 British cars by providing a measure of compensation equivalent to three million tonnes of CO_2. When the team from Channel Four's *Despatches* programme travelled to Mexico, however, they discovered the actual number was closer to 2,500 cars or just one third of 1 per cent of BP's impressive headline-grabbing figure. By the same token any supposed benefit accruing from the G8's new Cape Town roofscape is clearly temporary, and will disappear entirely when the shanty-town is dismantled and the residents are forced to move on. As for Britain's most popular carbon offset schemes – those for airline flights – these make no distinction whatever between local and long-haul journeys but simply charge a flat fee of around £10 per flight, which they say they use to plant a single tree.

One tree, clearly, is nowhere near enough – but then a major part of the difficulty here is in calculating precise carbon footprints, or if you prefer in seeing the wood for the trees, and what price you put on carbon in the first place. It clearly has a cost, it always has even if until recently we haven't had to worry about it, but serious estimates for this currently range from a rather modest-sounding $2 per tonne to a staggering $85, which as Britain's *Sunday Times* put it in a 'How the World Works' supplement in November 2007, 'would transform and probably paralyse the world economy at a stroke'.

Having someone diet for you

Part of the problem is that so far there is no accepted, independently verified formula for calculating precisely levels of carbon emissions. As a result airlines are free to take no account whatever of the so-called multiplier effect of releasing greenhouse gases at high altitude instead of at ground level, even though it is widely accepted that the impact on the environment of doing this is far larger.

Furthermore none of the schemes are independently rated or certified. Nor, even though the scope for fraud cannot be anything but very considerable, is even the best of them subject to any kind of quality audit or standardisation or required to operate within the terms of something like the UK's kite-marking scheme, which was conceived to advise consumers which operators (if any) are reliable and trustworthy.

There are some slightly more philosophical issues here too, so that Friends of the Earth is by no means alone in its belief that attempting to offset carbon in this way is a bit like announcing that you are going on a diet and then paying someone else to eat less on your behalf. In a sense the message broadcast by the aforementioned G8 scheme is akin to suggesting that it is perfectly OK for the developed world to carry on burning airline fuel like there's no tomorrow because South Africa's poor are saving on heating. Even more fundamentally there is also something rather questionable about the suggested equivalency between, on the one hand, some very real carbon from your car exhaust or one of the G8's private jets and on the other some theoretical carbon which may, after many years of growth, be temporarily stored in a forest somewhere far away.

Slow growth is no growth

Trees, after all, even mighty English oaks, do not live forever. Forest systems around the world operate on a cycle of many decades or centuries, rather than annually or over a few years as would be the case with conventional cultivated crops and many other forms of vegetation. Most significantly, because trees can take decades to reach maturity, they take many years to absorb the amount of carbon being offset. This means would-be offsetters have to consider two other questions: what's the fate of all the additional CO_2 they are producing in the meantime; and in a worse-case scenario what happens if the particular sapling they think they are buying dies young? Literally millions do every year, while millions more will be thinned out and burned as part of normal forestry management long before they have matured and absorbed any meaningful levels of carbon.

It is important to ask what happens to the carbon when this happens, what happens to the trees when they do grow old and die (or are cut down and harvested) and perhaps even more fundamentally whether any of the trees in question would have been planted anyway. If you plant your own tree you can be sure that it has been planted, and reasonably sure that it wouldn't have been planted anyway. But many commercial plantations in the UK, even those which are the recipients of offset monies of the sort described, were already planned and amply funded years ago by the Forestry

Commission. As a result only a very tiny fraction of their overall funding comes from the carbon offset companies.

Fans of the Rolling Stones, for example, are known to have paid out many thousands of pounds in a bid to offset the environmental impact of the group's 2003 tour in this way. They did this in the belief that the money raised – £8.50 per head – would be spent planting oak, birch and rowan on the Isle of Skye's extensive Orbost estate. In fact what the fans were paying for wasn't additional trees at all, reported Scotland's *Sunday Herald,* but rather what are called the 'carbon rights' to existing trees, specifically to some of the 128,000 saplings which were already scheduled to be planted using existing government grants. (Nor may it surprise you to discover that of the £10 payable for each tree in a similar scheme at the Glastonbury Festival as little as 54p actually reached the forest.)

Volvo similarly claimed it was 'creating woodlands' at Orbost – but actually, as the estate land is owned by the Government's own Highlands and Islands Enterprise, any extra carbon-offset revenues raised by Volvo merely plugged a temporary gap in HIE's finances and paid for the laying and refurbishment of a footpath. Little wonder then that a complaint about this and other schemes was eventually lodged with the Trading Standards authorities in London, or that Friends of the Earth Scotland called for this kind of carbon offsetting business to be regulated officially.

Perhaps even more significantly, though, many other critics of such schemes recognised that the whole issue of offsetting is really just a distraction from a problem people find too uncomfortable to address properly. In their hearts Rolling Stones fans, Volvo buyers and others, if they stop to think about it, could probably work out for themselves that if something is as easy as writing out a cheque for a tenner it probably isn't a viable way of tackling climate change. They probably know too that doing this sort of thing is easier (and a lot less painful) than making the very real lifestyle changes which would be necessary if they were really determined to cut their energy consumption – even if, however vaguely, they persist in their belief that there is some residual benefit in buying the 'carbon rights' to a pre-existing tree.

Where there's muck there's money

The concept of carbon trading – whereby big polluters and the largest individual carbon emitters have to buy permits for their extra emissions from smaller emitters who have surplus permits to sell – is similarly flawed, and as a consequence already largely discredited even though no actual scheme is up and fully running.

Here too there is the philosophical problem of allowing polluting companies to buy their way out of the problem rather than seeking to fix it by reducing their emissions. It has also become clear that some aspects of carbon trading will actually serve merely to increase the amount of CO_2 being emitted, for example where European countries choose to meet their obligations under the Kyoto protocol by sponsoring, say, a biofuels project in Indonesia, which in order to function will need to clear thousands of hectares of rainforest – or, worse still, peatbog – in order to have somewhere to grow the appropriate fuel-producing crops.

For example, it has been estimated by the charity Wetlands International that the destruction of peatland forest in south-east Asia alone – 130,000 square kilometres of it at the time of writing, most of it in Indonesia – has led to an average of 2 gigatonnes of CO_2 being released into the atmosphere every year. That's equivalent to almost a tenth of total global emissions from burning fossil fuels, thereby already negating many carbon-trading initiatives before these have even been put into operation.

Closer to home there is another concern. As we have just seen with carbon offsetting, the industry that has so

quickly grown up around the EU's own much-vaunted
Emissions Trading Scheme is certainly wide open to fraud in
the future. Because of this and other considerations many such
schemes begin to look less like a solution to the problem and
rather more like yet another business opportunity for those big
companies that are able to find a route through its many
loopholes. Certainly the BBC's *File on Four* programme of
10 June 2007 was unimpressed with the test phase, and found
little to say in favour of carbon-trading generally after
examining the effects of a large-scale trial scheme which
involved several leading UK power generators.

Most damagingly, the programme noted that because
carbon caps on individual companies had been set too high
from the word go – a product not so much of corruption, it
seems, but rather because the EU is inherently susceptible to
effective corporate lobbying – British power stations taking
part in the scheme were not just able to switch from 'clean' gas
to 'dirty' coal during the test phase in 2005 but still make
money while doing this. In fact in a single year their owners
between them scooped an estimated £1.2 billion in windfall
profits. Shell similarly reported a £49.9 million profit and BP
£43.1 million, much of it from selling their unused allocations
or permits.

Clearly the winners in this particular scheme clearly
won simply because they were able to sell carbon credits that
had been handed to them free of charge, instead of being
required to buy additional credits in order to offset the
pollution for which they are undoubtedly responsible.

Worse still is that a full third of the profits accruing to
the power station operators in this particular example were
actually paid for by domestic customers – yet another example
of a genuine stealth tax well beyond the inventive capabilities
of any satirist.

Noting a few months later that these new carbon
markets are simply too susceptible to 'fraud and inaction',

Derek Wall of Britain's Green Party called instead for a 'a new international agreement . . . to make actual cuts in CO_2 and other climate change gases, rather than providing a means for financial institutions to profit from ever more complex transactions'. In the present climate, however, such an agreement seems unlikely.

Chapter 6
THE END OF CHEAP ENERGY

There's no easy oil left. The only barrels now are going to be the tough barrels.

With our having consumed something like a trillion barrels of oil over the last one and a half centuries – and over that same period (in what is after all in geological terms barely the blink of an eye) released all the associated CO_2 – a partial solution to climate change, at least in one sense, will become apparent when the world's supplies of fossil fuels finally run out, one by one.

Once that happens and there is nothing left to burn, we will simply be unable to continue producing such large volumes of CO_2 and over time levels of the gas in the atmosphere should begin to scale back. That, however, is just about the only upside to this particular scenario, for if you do the sums it quickly becomes apparent that man's future – very much like the present, only more so – is going to be inextricably tied up with an ongoing struggle for energy, food and water.

All three, after all, are key resources, the production and consumption of which will need to be even more strictly marshalled and controlled than now – rationed is hardly an overstatement – if a world population of six billion and rising is to be reasonably provided for. Furthermore, even the simplest calculations show this to be the case regardless of how rapid or otherwise climate change turns out to be – and regardless of how that change finally manifests itself.

Global reserves already insufficient to meet demand

We already know, for example, that the earth's fossil fuel reserves are going to prove wholly insufficient to meet global demand – and that they will be exhausted within a relatively short space of time. We know too that – even without additional pressures such as climate change and a population explosion in the developing world – we are already consuming many other scarce yet increasingly essential minerals at a rate which is unsustainable. What is only now becoming apparent, however, is that many of the minerals already falling into short supply are the ones that have been driving the new technology forward. Precisely the ones, that is, that will be needed to make possible the new technologies which it was hoped would help us conceive and implement proper, workable, long-term and sustainable solutions to the problems of energy generation in the future.

Understandably, optimists and a dwindling band of climate-change sceptics insist the picture is not as black as it is painted, because the full extent of Earth's natural resources is still something of an unknown quantity. This is true: without doubt new reserves will be found, while precise figures of known reserves are kept a closely guarded secret by those commercial mining companies which rely on their extraction and sale for their profits. Many of the same people also argue that the nature of scientific endeavour means the coming years are bound to reveal other new technologies of a sort we cannot presently conceive, and to throw up unexpected answers to at least some of the more pressing environmental questions. This is also certain to be true.

The fact remains, however, that the problems facing mankind are real, immediate, immense and becoming ever more pressing. It is true that the debate about precisely when the oil will run out is still ongoing – but no-one seriously imagines that it will not run out, that major price rises of the sort now being seen are not going to become the norm, or that any of the alternatives so far proposed constitute anything approaching a workable solution. Similarly, just because political considerations and commercial confidentiality prevent us from knowing just how much of a given supply remains, this is hardly a good basis on which to assume the supply is infinite.

Big Three becomes Little Three

The problem of resource depletion affects many different natural substances, but not unnaturally the main focus has for twenty years or more been on the so-called Big Three of oil, coal and natural gas. These attract both the most media attention and the most public comment because we all rely on them to an overwhelming degree and in ways we can all understand. Also, of course, because in the climate-change debate the conversion of these fossil fuels into planet-warming CO_2 means that they and their applications are coming under scrutiny as never before. And, finally, the fact that all three are strictly finite resources which are being consumed faster than ever before serves to focus public attention on what is to be done once the supply finally dwindles to zero.

Fortunately this last problem is now, at last, being recognised as imminent and acute. Acute because modern life without any one of these non-renewable fuels – let alone all three of them – is hard to imagine. And imminent because all three are being consumed in such volumes, and at such a speed compared with the millions of years it took to form their vast reserves, that it is inconceivable that their practical exhaustion can be very far into the future.

Coal, of course, has the longest history, having been an essential source of power since the eighteenth century and Britain's Industrial Revolution - although it is only in the last 100 years or so that such vast quantities of it have been extracted from underground workings. Oil and gas have also proved their usefulness over many decades, most obviously in the case of oil to provide the power for internal combustion engines and in the case of gas to provide the fuel for a new generation of cleaner, more efficient power stations. Of course, albeit much less visibly, oil has also provided a vital feedstock for many of our largest and most important industries, such as those involved in chemical and plastics manufacture, as well as others playing a major role in agricultural supply.

But unfortunately, and regardless of the size of their known and unknown reserves, supplies of all three are signally failing to keep pace with demand. It is also the case that – whether one alights on nuclear, solar, wind, wave or biomass alternatives – there is as yet absolutely nothing out there which can even come close to matching fossil fuels in terms of their energy-storage value, their efficiency or their final cost to the consumer.

At the same time the present infrastructure is entirely geared towards petroleum, so that while renewable energy probably has a niche role to play in local applications it will never be sufficient to power a global economy. Renewable energy will not, for example, replace a fleet of vehicles. Instead, the existence and bold claims of the renewable energy lobby have chiefly served to obscure the real issue, which is our urgent need to slash consumption to an absolute minimum.

This being so, debate and discussion continue to centre on the precise size of the remaining fossil fuel reserves, and more pointedly on the question of how soon they will run out. Here at least, and at last, there is something approaching broad agreement that the timescale is by any sensible reckoning very short indeed.

How soon is soon?

Despite the occasional positive headline it is generally acknowledged that the good years of plentiful and cheap fuel are already behind us – even if for most of that time people were generally unaware that they were living in a relatively low-cost economy.

It is known too that the days of the really big oil finds are also over; that while the early oilmen had to rely on relatively primitive exploration techniques, the peak year for discoveries of major oil fields could have been as long ago as 1930 in the US and 1962 for the rest of the world. Of course new finds will continue be made, but it is worth reflecting on the fact that in the last two decades or so fewer than a handful of major fields have been discovered which comprise more than a billion barrels apiece – two in Brazil, and one each in Colombia and Norway, none of which produces more than 200,000 barrels per day. It is also sobering to think that around 80 per cent of the oil consumed in, say, 1995 was actually discovered many years before that, and certainly before 1973.

No less worryingly – and even without taking into account the spiralling energy requirements of the rapidly developing economies of India and China – is that demand for all fossil fuels continues to increase apace. Currently it is doing this to such an extent that, while new discoveries of oil between 1990 and 2000 totalled an initially healthy-sounding forty-two billion barrels, consumption over that same period exceeded that same amount by a factor of almost six. At current values, says the Association for the Study of Peak Oil, we typically consume in excess of four barrels for every new one we find·, meaning that we are not just outstripping our own ability to find more oil but also the earth's ability to make it. (The approximately twenty-nine billion barrels mankind collectively manages to consume in a given year is roughly equivalent to the amount of oil which would have taken a staggering nine million years to accumulate.)

Set against these observations it is scarcely surprising to hear the likes of Goldman Sachs noting that 'the great merger mania [seen in the oil industry] is nothing more than a scaling down of a dying industry in recognition that 90 per cent of global conventional oil has already been found'. This particular remark was made several years ago, and today is really only worth repeating because of its baldness and the implications it contains. Additionally, there has been nothing to suggest that a change of view might now be in order.

Don't be fooled by the figures

Admittedly from time to time new or revised estimates emerge which might suggest some small cause for optimism – but all too often a simple explanation can be found for this.

For example, the industry body OPEC takes the size of a country's reserves into account when fixing annual production quotas. (In other words, the more oil you have or say you have in the ground the more of it you are allowed to sell.) These same reserves can be used as collateral for loans, giving producer-nations yet another reason to publish over-optimistic figures.

An example of this might be the money lent to Mexico by the US when the peso collapsed in December 1994. With its currency losing 35 per cent of its value, Mexico's fiscal reserves fell like a stone from \$29 billion to \$5 billion. This prompted President Clinton to agree to a \$50 billion Emergency Stabilisation Package in order to prevent a national economy collapsing on his southern border. The collateral for this was a pledge of revenues from Mexico's future petroleum exports, so it was in nobody's interests to do anything but overstate the likely size of these revenues – or the likelihood that they would continue flowing for many years.

Set this kind of statistical manipulation against an authoritative report from an organisation such as the United States Energy Information Administration (EIA) suggesting that over the next twenty years demands in the US alone are predicted to rise by 33 per cent for oil – and perhaps twice that for gas – and suddenly one begins to see why, in the EIA's own words, 'America faces a *major* energy supply crisis over the next two decades.'

Crunch-time: costing the earth

The truth, however, is that we *all* do, which is why as long ago as 1999 US Vice-President Dick Cheney, wearing the CEO's hat at Halliburton, the world's largest oil services company, was putting the case even more forcefully. At that time, in a speech to the International Petroleum Institute in London, he was able to demonstrate that even the most modest estimates of just 2 per cent annual growth in global oil demand, and a similarly conservative estimate of decline in production of just 3 per cent, are together sufficient to indicate that 'by 2010 we will need on the order of an additional fifty million barrels a day, equivalent to more than six Saudi Arabias'.

Of course, with barely two years to go, nothing of this magnitude has so far materialised, and nor is it expected to by those in the industry. J. Robinson West, chairman of oil industry consulting firm PFC, put it rather well in the *New*

York Times when he said, in October 2007, 'There are no easy barrels left. The only barrels are going to be the tough barrels' – hence, he noted, the tripling of oil discovery and development costs between 1999 and 2006.

Claude Mandil, head of the International Energy Agency, the intergovernmental energy watchdog, has also expressed concerns about future security of supply, noting that many important geological reserves 'are in regions closed to foreign investors, like the Middle East, or increasingly hostile, like Russia or Venezuela.' As a result, he says, oil majors such as Exxon, Shell and BP 'have a serious problem: their targets are getting scarce'. In fact there's a mass of contemporary geological knowledge, together with backdated reserve figures and production histories, to suggest that the decline of all liquid hydrocarbon production is a mere two or three years away from today.

At that point supply will drop away markedly and rapidly, and the price will rise accordingly – an especially daunting thought when one considers that nearly all of western economic development and prosperity since the nineteenth century has been established absolutely on the foundation of affordable, abundant energy.

This means of course that it's not going to be just private motoring which will be affected – although at times this seems to be the public and the media's chief concern. Instead, almost every aspect of modern life that we take for granted will be affected negatively. Even the production of affordable food depends increasingly on industrial-scale agriculture – something which has a huge appetite for energy, which is needed to manufacture artificial fertilisers, fungicides and pesticides, not to mention to build and power all the industrial machinery needed for planting, cultivating, harvesting, processing, packaging and transport.

The role of coal

Albeit in only one regard, coal might provide part of the answer since there are still in excess of 909 billion tonnes of proven reserves worldwide. According to the World Coal Organisation's own figures this sort of quantity would be sufficient to last approximately 150 years. (This is at current rates of usage, although consumption of coal is rising fast even in the UK, and globally is expected to rocket by more than 70 per cent by 2030.) Also in its favour is the fact that its reserves are located in more than seventy different countries. That might reduce coal's exposure to political pressures of the sort which continue to dog oil (and increasingly gas supplies as well) although the reality is that up to 90 per cent of the reserves are concentrated in just six countries.

Additionally, and especially in a book primarily concerned with climate change, the downsides of coal as a source of heat and power can scarcely be ignored. Per unit of energy released it produces a staggering 50 per cent more CO_2 than natural gas, and as if that weren't bad enough, tonne for tonne it is a far less efficient source of power than either oil or gas. So far at least it is also significantly more limited in its range of potential applications.

Not, mind you, that any of this has prevented the promotion of some slightly desperate and extremely heavily subsidised coal-to-liquid-fuel projects of the sort currently being driven by the US House Committee on Energy and indeed by US presidential hopeful Barack Obama. The latter is clearly keen to use the estimated 100 billion tons of coal thought to be lying beneath his home state of Illinois for transportation: in Washington he's on record as saying the people he meets back home 'would rather fill their cars with fuel made from coal reserves in southern Illinois than with fuel made from crude reserves in Saudi Arabia'. But against this is the opposition of many independent specialists in the field, experts who in particular warn that a failure to devise and use effective carbon capture and sequestration technologies (assuming these even exist) mean such initiatives as Obama's will likely lead to an increase in global-warming emissions rather than a decrease.

Because of this they suggest instead that research spending should be directed towards plug-in and hybrid vehicles of a sort designed to run using electricity produced externally by cleaner, reduced-carbon sources. According to the Carnegie Mellon Electricity Industry Center, the most pressing goals of America's national energy independence – supply security and reduced greenhouse gas emissions – could both be achieved in this way and at much lower cost. Later in this chapter, however, we shall see that plug-ins and hybrids have their own problems when it comes to the depletion of other similarly scarce natural resources.

As for gas, it too has its supporters. Certainly it is cleaner than coal – most alternatives are – as well as far more versatile and a much more efficient source of power. Its reserves are also reasonably substantial, but they are also more difficult to assess than oil. Gas is particularly susceptible to a variety of different economic factors and, like oil (as we are now seeing with Gazprom, in which the Kremlin still has a controlling majority stake), to political manipulation of a most unwelcome kind.

Unfortunately its uniquely high mobility also gives rise to what are known as 'gas bubbles', when the market becomes flooded, while the fact that its supply is largely tied to existing pipeline networks – whereas oil can be shipped in from anywhere by tankers – means that certain countries (the US is the most obvious example) are likely to run into supply problems a lot faster than others. Not that the rest of us will in any way escape the inevitable, however, for even if energy consumption continues at only its present pace – something which is of course wholly unlikely – supplies will not last long. In fact if natural gas were to replace oil, and if the current level of economic activity were maintained, it is thought that global natural gas supplies would be exhausted within thirty to thirty-five years.

This in turn means that any attempt to use natural gas as the primary substitute for oil will at the very best provide only a temporary solution. More likely is that a switch from oil to gas would be a distraction, a major readjustment which accomplishes little beyond the squandering of both human and industrial resources on the production of vehicles, power stations and manufacturing processes which would in turn become obsolete. Far better, say many observers, that these same resources and our remaining financial reserves be used to find and implement more innovative and more sustainable solutions.

The impact of developing economies

Not that any of this has prevented some observers from claiming to have found some small cause for optimism. This they have done after noting that over the last two decades economic growth in several advanced economies – including the US and Germany, Japan and the Netherlands – has been at a faster rate than that at which they are consuming natural resources.

From this basis it is argued that, if such a trend can be encouraged and intensified, future economic growth can be divorced from continued reliance on a dwindling supply of natural resources and thereby can continue unabated. To this end (having calculated that at present it takes approximately 300 kilograms of natural resources to generate just $100 of income), members of the Organisation for Economic Co-Operation and Development have agreed a preliminary target of reducing this to just 30 kilograms over the next several decades.

But unfortunately, like so much in this debate, many of the observable gains are just illusory. In large part this is because the problem of resource depletion has become yet another aspect of modern commercial life which has effectively been 'off-shored' rather than actually resolved.

These days in the West we manufacture much less of what we consume, and prefer instead to concentrate on service industries and the so-called 'knowledge economy'. By their very nature these use less in the way of natural resources – yet as we continue to consume more than ever the same resources are still being used up, but being used up elsewhere on our behalf. In fact they are being used up more quickly than before too, the difference being that now, as opposed to fifteen or twenty years ago, the rate at which they are being depleted appears on the exporters' balance sheets – increasingly in countries such as China and India, which already account for more than 45 per cent of all the coal burned – rather than on our own.

In other words, and looking at the global picture, the response to what has become a serious and looming crisis has been to press on regardless. That's true even though the more we dig and drill the more CO_2 we will produce, and against a background where demand for energy is set to grow by another 50 per cent by 2030.

Chapter 7
SAVING ENERGY WON'T SAVE US

The only way to accomplish any meaningful savings in personal emissions is to reduce one's consumption, use the money thus saved to buy exploitable sources of fossil fuel – and then leave them in the ground.

Whether one is discussing oil, coal or gas, and regardless of the actual use to which they are put, fossil fuels constitute a relatively simple form of economic commodity. As such they obey the classic model of supply and demand, and of price equilibrium. This means that as the supply of the fuel in question increases – and in this case we're really talking about the price of energy as a whole – the cost to the user comes down, thereby leading inexorably to a proportionate increase in demand. This remains true regardless of any specific political situation and, particularly in the more urgent case of oil, it can be seen to be true regardless of whether production peaked years ago, is peaking now or will peak shortly as global reserves dwindle to zero.

The equilibrium is therefore the point at which demand equals supply, although in this particular case such a point is never reached as demand for energy is always certain to outstrip supply. Mostly this is because whatever new technologies are devised and whatever new reserves are found, there will always be new or extra demand to soak it up. Even if this additional demand does not manifest itself in the prosperous West, it will certainly be evident in the developing world, where, even now, most of the population of Africa still lacks access to a reliable supply of domestic electricity. Just as certainly, this demand will be fed by the energy companies for as long as they are physically able to obtain the necessary raw materials.

Super profits, super problems

With fossil fuels in particular, there are two other important considerations to take into account when looking at the issues of supply and demand. The first is that we have in a sense already consumed all the oil, coal and gas reserves the planet has to offer – in that we are consuming them as fast as the energy companies and producer nations can wrest supplies from the ground.

The second is that the price we as individuals pay for power (which industry pays likewise) has very little, almost nothing, to do with the cost of production, because oil, coal and gas are in reality extremely cheap to produce and distribute. This situation gives rise to so-called 'super profits', vast revenues which have traditionally been split unequally between the big oil multinationals, the governments of the producer nations (or in some cases their ruling families) and the governments of consuming countries. Although the precise nature of the division between these three entities varies from country to country and from time to time, it is achieved in the same way through various different royalties and taxes.

This in turn leads to something of a contradiction in the usual notion of smooth supply and demand, which is that in circumstances where the supply is inelastic the price to the end user should be highly volatile. With oil and gas this is rarely the case, however, although this can be explained by looking at the whole of the energy market, and not just that for oil or gas in isolation.

Once this is done it quickly becomes apparent that it is the more traditional fossil fuel – i.e. coal – which supplies the necessary elasticity in the market. In other words it is coal which over the long term has functioned to keep the price relatively steady, although this elasticity has to some extent been evaporating in recent years – not least because the cost of extraction for oil and gas has remained very low relative to that for coal, especially where the latter is deep-mined. As a result of this oil and gas have been slowly displacing coal over the last thirty to forty years, mines have been closing down – and not just in the West – and far smaller volumes of deep-mined coal have been extracted than at any time in the recent past.

Now, however, with supply of the two more cheaply extracted fuels falling far behind demand and prices rising fast, coal is proving to be resurgent once again. This is especially true at those locations where it can be obtained more cheaply from open-cast mines – and where in recent months we have seen a dramatic increase in accidents, suggesting that safety is being compromised in the race to raise productivity. China, for example, responsible for 36.1 per cent of global coal consumption, now gets nearly all its power from coal, as does

Australia – albeit on a far smaller scale. Similarly another rapidly expanding economy, namely India's, while still behind the US, is now using more and more coal, with predictably disastrous results for the environment and some very serious implications when it comes to the pressures of climate change.

Why using less energy won't cut emissions

It sounds logical at first, but unfortunately cutting back is not that simple. Power stations cannot simply be switched on and off as demand rises and falls throughout the course of a day. Instead the infrastructure of a national power network or grid has to be kept operational around the clock, and as often as not much of the equipment which makes up that network has been designed to run at a set maximum in order to operate at optimum efficiency.

Besides this fairly fundamental limiting factor there is something else to consider too. Reducing CO_2 might well have an impact on climate change eventually, but saving energy is not necessarily the same as reducing CO_2. In fact according to many economists energy reduction is a major contributor to economic growth: householders saving energy (and so money) by buying a newer, smaller, cleaner car when the time comes, or by choosing domestic holiday destinations over long-haul ones, may well find themselves better off by the year-end.

Chances are that they will spend the money they save, and the chances are also that whatever they buy will involve some additional expenditure of energy. The only way to accomplish any meaningful savings in personal emissions would be to reduce one's consumption to live a genuinely low-carbon life – a very radical change which is not easily accomplished – using the money thus saved to buy exploitable sources of fossil fuel and then leave them in the ground. The fact that no-one does this is called the energy-rebound effect, and of course it affects businesses too and on a much larger scale. Companies may, for example, institute an energy-saving

programme, thereby reducing their fuel bills – and then use
the resulting rise in profits to build another factory producing
yet more CO_2.

Finally, and even if this energy rebound effect doesn't
play a major role, there is another problem to contend with.
The fact is that were you as an individual to reduce your
demand and to do so on an ongoing basis – and were
sufficient numbers of similarly public-spirited people were to
follow suit – the resulting drop in local demand would simply
cause the price of energy to fall. And unfortunately once that
happens any other would-be user who has so far been
unwilling or unable to consume as much energy as he might
like to (on account of the cost) suddenly finds himself in a
position to do so thanks to this new, lower tariff.

How the market promotes pollution

This, after all, is one of the very foundations of our modern
economy, the way in which all competitive markets function, and
this default model has been engineered over centuries to maximise
both production and consumption. Essentially this is achieved by
increasing overall efficiency so that prices can be lowered in order
to bring ever larger numbers of consumers into the market. This is
a model in which even much poorer, less developed countries
willingly conspire – clearly doing so in the hope and belief that in
time the same process will enable their own quality of life to rise to
meet that of the developed world.

Put this way, such a scenario is morally desirable too, in
that it offers the world's poor a chance to carve themselves a larger
slice of the energy pie – albeit using technology likely to be less
efficient and more polluting than that employed in the West. Its
other unfortunate corollary is that the total amount of energy
consumed will as a result remain more or less stable. Yet having
enjoyed relatively cheap energy (and its obvious commercial
benefits) over such a long period we in the first world are hardly in
a position to deny those same benefits to those in the third.

Neither, for the same reasons, can either central or regional governments take effective steps to cut emissions by ensuring a greater proportion of the population use energy more efficiently – although this has not prevented many of them from attempting to do so.

In fact successive administrations around the world have in recent years introduced all sorts of different initiatives, from gentle persuasion via public-information advertising, through to personal carbon calculators on the internet, and occasionally even modest incentives to encourage 'greener' behaviour. Unfortunately and despite their worthy intentions most of these efforts are at best marginal or somewhat misguided, besides which, when it comes to persuading or cajoling the general public to act socially or 'responsibly', it is rarely enough merely to attempt to shift opinion and then hope for the best.

Recognising this, these same authorities usually find they are eventually forced to consider wielding a stick as well as the odd carrot, most commonly doing this by introducing carbon taxes of one sort or another. These are configured in such a way as to ensure that the more fuel an individual or company uses the higher the price they have to pay for it – the hope being, of course, that such taxes will eventually force consumers and businesses to reduce their energy consumption (and so their emissions) in an attempt to cut their annual bills.

Unfortunately the revenue thus generated is rarely if ever hypothecated, which is to say set aside for specifically green projects. More commonly it is simply thrown into the Exchequer's general pot and lost to view – thereby boosting public cynicism still further and leading increasing numbers of taxpayers to the conclusion that the whole thing is simply just another stealth tax masquerading as a green one.

Their efficacy is unsurprisingly poor as a result – though even were it not this sort of tax would simply depress demand at a regional or national level, thereby leading once again to a lowering of the price internationally. This in turn would allow poorer nations to increase their own consumption – more Africans switching on more lights – while at the same time enabling any country which had decided not to impose any taxes of this sort to jump in, steal the benefit for itself, and in this way increase its own economic competitiveness at the expense of those nations which have chosen to take a more environmentally sensitive line.

Either way, the amount of energy used would at best remain more or less stable, and as developing nations invariably rely on less efficient plant and older machinery than we have in the West, as mentioned above, the amount of carbon emitted would in all likelihood go up, not down.

What if everyone acted together?

Of course it is reasonable at this point to ask what would happen if every government in the world decided to act in concert for the good of mankind? What if every country on the planet agreed to introduce a workable and comprehensive carbon tax regime designed to reduce consumption and so CO_2 emissions? If this was done then the problems of substitution and competitive advantage would be ruled out immediately – wouldn't they?

Sadly not. Even were such a thing possible, all it would do is transfer wealth from the producer nations to the consumer nations by substantially reducing the opportunity for the governments and ruling elites in the former to extract royalties from the oil majors prospecting on their territories. The market would simply adjust, in other words, or rather it would were the world's governing authorities all to agree to act in concert – which, of course, they are never going to do.

There are many different reasons for this, not least the natural but very considerable tension which exists between producer nations and consumer nations. There is, after all, no rational or correct basis for dividing the aforementioned super profits between the two sides. On the contrary: all we have currently is a pragmatic solution regarding the historical levels of tax and royalties, although even this shows a very wide divergence across the market. In a country such as Libya, for example, oil production is nationalised so that the state keeps all the royalties; meanwhile in the US oil extraction is entirely in private or corporate hands and there is virtually no tax levied on oil production.

The situation is indeed much the same in the consumer countries, so that in North America taxes on gasoline are extremely low. China too, clearly keen for quick and considerable economic growth, charges very low taxes on all forms of energy usage, whereas in the UK and most of Europe consumers have grown used to much higher levels of tax on oil, gas and so forth.

Such a situation leads inexorably to the next great disincentive to any harmonisation of fuel tax regimes – which is that any country with a very low tax on fossil fuels finds itself with an immediate and substantial economic advantage over its high-tax competitors. This is evident right across the spectrum, from manufacturing through transport to agriculture and construction, as well as when it comes to the cost to citizens of heating their homes and fuelling their cars.

Taxes don't work anyway

It's also worth considering the testimony of R. James Woolsey to the US House of Representatives Select Committee on Energy Independence and Global Warming in April 2007. In particular he noted that 'oil use in transportation is only lightly affected by . . . carbon taxes or carbon cap-and-trade systems'. This is because 'an increase in price of many dollars per ton of CO_2 will have only pennies' worth of effect in the price of gasoline. So while such methods of limiting emissions from coal combustion have much to commend them, they have little to do with reducing the over 40 per cent of CO_2 emissions that come from oil, especially in its transportation uses.'

Carbon taxes, in other words, make no sense whatever – especially for a government intent on securing full or near-full employment. They are also extremely difficult to sell to a population which is keen to see economic growth and greater individual wealth, particularly (in the case of the likes of India and China) when that population can see that the West built its wealth on the back of green-free energy policies – so why should not they?

In addition, if somewhat perversely, in a would-be service economy such as Great Britain's, high fuel taxes also provide a very strong incentive to purchase manufactured goods from these same low tax countries rather than making them at home. Indeed, by acting in this way many of the more developed countries of the world could be said to be importing low-tax energy in the form of finished goods – although this peculiarity hasn't prevented member states of the EU from being somewhat boastful about their falling carbon emissions in recent years.

It can of course be seen that this is a quite specious argument, since any so-called gains are almost entirely down to the 'off-shoring' of manufacturing in this way: the emissions are still produced, just not at home. Worse still, perhaps – and leaving aside the moral problems of child labour, extraordinarily low rates of pay and so on – this manufacturing is being delegated to countries where energy production and energy conversion is far less efficient and far more polluting than it would be in, say, Europe, Canada or the US.

China, as a consequence, accounts for just 5 per cent of the world's total economic activity yet consumes more than 10 per cent of its energy, up to 20 per cent of its cement, a horribly energy-intensive commodity, and together with India accounts for nearly half of the growth in the amount of coal which is being burned. Politicians and civil servants appear to be too dumb to see this, however, or to recognise that what they term free trade is in reality little more than an enormous tax-avoidance scheme.

So what if we just stop digging?

One solution to the whole supply-and-demand problem, and one that could have a substantial impact on the problem of climate change were it to be implemented soon enough, is for mankind to simply stop digging and drilling. To restrict the world's supply of fossil fuels, that is, and to do it on the sensible grounds that if we were to take less oil, gas and coal out of the ground then by definition – in time – there would be less CO_2 released into the atmosphere to cause a problem.

Restricting supply in this way would clearly raise the price of energy, thereby encouraging more sensible, more effective use of the available supply. In this way it would also and more significantly incentivise both individuals and industry to be more efficient in the way fossil fuels are consumed and to seek more viable alternatives to oil, coal and gas. From an economist's point of view restricting supply in this way would also be a foolproof method of ensuring that the energy, as it becomes much more expensive, is directed towards the very highest utility – in other words, to wherever mankind would obtain the most bang for his buck.

It makes good sense from a producer-country's point of view too, of course, since a hike in prices would effectively compensate producers for the much lower volumes being extracted and sold. The revenues accruing to these countries could even increase quite markedly, as happened in the 1973 oil crisis when a politically inspired Arab oil embargo in the wake of the war with Israel caused the price to rocket tenfold around the world.

In 1973, though, the unexpectedly sudden and precipitate price rise proved quite disastrous. Creating severe power shortages around the world, it led to dramatic inflation as the increased cost of transport fed directly into the price of literally all goods which were being transported – including, most damagingly, ordinary food. But more positively it also provided what the leading British historian Sir Martin Gilbert referred to as a useful 'spur to scientific development' – something which is badly needed at the present time.

Such a move now need not have such a damaging impact, not least because the volume of oil, coal and gas being extracted and refined would have been cut not for political reasons or to extort even greater profits but to save the planet (or more accurately man's place upon it). For the same reason one would like to think that we might avoid a damaging consumer backlash against the producer nations.

Limiting supply in this way would also be relatively easy to audit, since oil, coal and gas occur in only relatively few locations, all of which are well known. For the same reason policing the restrictions would be straightforward too, because supranational authorities would be able to ensure that individual supply quotas were adhered to by threatening to drop a single well-placed bomb on a renegade refinery, gas plant or oil well rather than having to consider engaging with and wiping out an entire army.

In fact the underlying logic of such an approach – or more pointedly the fact that no-one seems even to be considering it – makes one wonder whether there is a conspiracy of silence regarding the whole issue of global energy supply. Certainly it is not hard to see why the richer, more powerful net-importers of oil and gas might object to the plan; life for their countrymen would get harder and a lot more expensive, and of course were we to stop digging the effect would be to transfer wealth in the form of taxes and oil company profits from consumer countries to the producer ones.

It is true too that many other conflicts could potentially arise where an energy producer nation is also a large consumer. China, India and Russia, to name the big three, all have rapidly growing manufacturing bases which rely heavily on the availability of increasing amounts of economically priced fuel. This means that they have little, if any, incentive to stop digging. Eventually it all comes down to whether one man is willing and able to make the first dramatic move – and that man is the ruler of Saudi Arabia.

Arabia to the rescue?

What if a Saudi monarch decided unilaterally to try and save the planet? With such a large proportion of the world's production and reserves to call on – 262 billion barrels of oil, or about 25 per cent of the global total, and more than 6.5 trillion cubic metres of gas – he, and it is always going to be a he, is certainly uniquely placed in that his own individual preferences and actions could make an impact.

Against this it is true that 75 per cent of all government revenues in his country come from oil and gas. Then again, and as previously observed, cashflows into the country would be completely safeguarded by the price rise, which could not fail to follow a unilateral decision to stop drilling. Nor need the king be overly concerned about any unemployment which might result from a cut in production, since more than a third of the workforce in Saudi Arabia is currently foreign nationals rather than his own subjects.

In theory, then, His Majesty could announce that he is going to cut production by, say, 10 per cent compound, i.e. year on year, and seek to encourage other, smaller producers to declare a similar policy. (That said, if they chose not to follow his example it would not matter a great deal, since none has the capacity nor capability to compensate for even the partial loss of such a major supplier.)

Prices would of course rise steeply and rapidly. Having set an example in this way, His Majesty could then sit back and watch as consumption dropped off at a more or less equivalent rate. Before long emissions would fall even further as the technologically more advanced of the consuming nations became more imaginative, more efficient and even more effective in the ways in which they used the available energy.

As a consequence, over time, climate change would slow and the world would be saved. Or would it? Would other smaller producers follow suit? Would every consumer nation allow the king to act in this way? Does he even want to save the planet?

Well, that's a tough one. Clearly many smaller producer nations lack the wealth and the infrastructure which would in theory allow the House of Saud to act in such a strong and decisive manner. Some, of course, also lack the necessary political stability to risk taking such a politically hard and controversial line. At the same time the majority of big consumer countries are almost certain to be highly antagonistic to the idea: at least one of them is already seen in some quarters as supporting puppet governments in several major oil-producing countries precisely in order to ensure that these continue maximising production and that they, the Americans, enjoy first call on those supplies.

The fact remains, however, that we cannot be sure that the right kind of decisive action is being taken unless or until the volume of oil, coal and gas being extracted begins to fall. That in itself is easy enough to monitor – just don't hold your breath waiting for it to happen – and of course even then we would still have to wait for many years to see whether cutting production and consumption was actually enough to avert disaster. After all, as Roger Clark of GlobalWarmingResponse.

org put it in a letter to *New Scientist* back in August 2007, if mankind has thrown a hammer into the climate works, 'promising not to throw any more spanners in is scarcely going to mend the machine'.

Chapter 8

THE BIOFUELS FALLACY

If this is a horse race, we're betting on a donkey.

The theory behind biofuels as a more sustainable alternative is one which has been understood for many years, and much of the technology has been available for use for just as long. With very few exceptions, however – Brazil being the best known – it is only in the last decade or so that any serious progress has been made in terms of their actual substitution for more conventional fossil fuels.

Any such advances have largely been made as a result of US and EU government policies and subsidies rather than from changes in marketing or consumer demand. It has also to be said that in general these government initiatives find their origins not in environmental concerns but in the desire to improve the security of supply, by reducing consumers' dependence on fossil fuels where its control lies beyond their own national borders.

It is for this reason that in the US, for example, maize being grown for ethanol production has now become the single most subsidised crop in the entire country. With $51 billion being paid to growers in the decade before 2005, the stated intention is now for at least 20 per cent of all transportation fuel consumed in the country to be either ethanol or biodiesel within ten years. The EU has called for a smaller target of 10 per cent, with a number of incentives being introduced to this end.

These include the UK government's recent decision to scrap duty on the first 2,500 litres of 'home-brew', or biofuel derived from waste vegetable oil. (Although, just weeks later, a leaked briefing document appeared to suggest that the very same government was actively casting around for ways to avoid meeting these EU targets for renewable energy and to offload a larger proportion of the burden onto its fellow member states.)

Unsurprisingly, what amount almost to threats from the likes of Abdalla El-Badri, secretary-general to the Organisation of Petroleum Exporting Countries (OPEC) have provided another very strong incentive to seek alternatives. For example, in June 2007 the *Financial Times* reported him warning that 'western countries in their efforts to develop biofuels as an alternative energy source to combat climate change risked driving the price of oil through the roof'. It is also instructive to note that the biggest demand in America for these supposedly greener alternatives came not when Al Gore's campaigning environmental film went on general release but when the price of gasoline rose above $3 a gallon.

Whatever the impetus behind it, the uptake in the US has been remarkable. According to the Renewable Fuels Association in Washington DC, for example, ethanol production in the US effectively doubled to more than five billion gallons a year between 2001 and 2005. The number of plants producing such fuels also doubled over the same period to around 120. The Food and Agricultural Policy Research Institute similarly quotes a doubling of maize production for ethanol in 2007 – compared with the figure for just two years previously – while saying it expects a full 32 per cent of the total tonnage of the crop to be used for biofuels by 2009. For his part President Bush has called for the total volume of US biofuels to rise to an incredible 35 billion gallons by 2017, and for a full three-quarters of all imported oil to be replaced by biofuels by 2025.

Few countries can match this level of escalation, but around the world a number of different fuels have come into play in recent years – as described below. At first glance each looks like an attractive and sustainable alternative to oil, especially when their positive attributes are viewed against a background of the dwindling and clearly finite supply of the latter. It is also very much in their favour that most if not all can be homegrown, particularly when industry observers report an apparently increasing willingness of those who control the supply of more conventional fuel sources to flex their muscles for political and/or economic gain.

But unfortunately none of the biofuels, either separately or together, provides answers to the two most pressing questions. How do we avoid climate change, and what happens when the oil runs out? This chapter explains why they never will, although to do this we need first to take a look at what the leading biofuels are, and then to reveal the carbon-neutral con.

Biomass

In this context the term biomass refers to the use of organic material as a feedstock for a new generation of boilers and generators, units capable of providing heat or power (or a combination of the two) and specifically designed to burn straw, wood chips, animal waste, rice husks, even domestic sewage or food waste.

In its favour, and using relatively cheap and simple technology, biomass can therefore be drawn from a very wide variety of different sources. As a fuel source it could also in principle mitigate the problems of waste management as well as those of domestic fuel security and climate change. Very much on the debit side, however, is the problem of transportation, since the energy derived from biomass is typically around just half of that which could be extracted from a traditional fossil fuel, while the volume of the same feedstock is between 400 and 500 per cent greater than oil, liquefied gas or coal.

In practice what this means is that moving biomass around requires a lot of trucks. Worse still, these tend to be full rather than operating at their efficient weight limit, meaning that each truck is typically likely to be carrying just 8–10 tonnes of bulky material rather than the 38 or even 44 of denser material which the largest models are now technically and legally capable of doing. Unfortunately the low density of these sorts of materials also makes for relatively high handling costs at both ends of the process, where they are harvested and where they are fed into the boiler. This is not just because you need to move a much greater quantity but also because, compared with oil, gas or even coal, it is very hard to automate the handling process in order to make it faster and more efficient.

Finally there is the issue of moisture. Fossil fuels are effectively dry and stay dry even when it rains; biomass, on the other hand, has a tendency to absorb and store additional moisture at every possible opportunity, which means it burns less well and much less efficiently.

Biodiesel

Derived from a similarly wide range of different crops, including rape or canola, soybeans, palm oil, hemp and sunflower sources – as well as used chip shop oil and even waste animal fats – biodiesel is probably the best alternative fuel in terms of its overall efficiency. The seeds from which it is generally derived on a commercial scale are dry and relatively dense, making them cheaper and easier to handle, while the relevant plant species which supply the seeds are extremely widely distributed throughout the world and found in many different climatic zones.

Equally fortunately, and regardless of which particular source of seeds is chosen, the means of producing the actual fuel is also extremely straightforward. To do it the seeds are simply crushed – leaving the waste materials to be used for animal feed or indeed as biomass – after which simple and basic chemical processes can be employed to convert the resulting oil into something which burns much like conventional diesel fuel and can be used in many different applications.

Unfortunately, however, very large-scale manufacture of biodiesel in south-east Asia – particularly involving the production of palm oil in Malaysia and Indonesia – is having a highly deleterious effect on the environment: it is increasingly being linked to massive deforestation.

Here great swathes of ecologically valuable forest land are being cleared each year, and the timber burned in the open, releasing many thousands of tonnes of CO_2, in order to make way for large, industrial cultivation of more suitable feedstock species. Indonesia, indeed, is now the third largest emitter of greenhouse gases after the US and China, in part because it is destroying its rainforests but more significantly because it is draining its peat bogs to free up land for agriculture, thereby releasing even more CO_2 acre for acre than many western cities.

Generally accepted estimates indicate that draining an acre of bog will actually lead to the emission of around thirty times as much carbon as would be saved by burning the biofuel produced at the same location in place of conventional fossil fuel. Put another way, were Indonesia to drain the remainder of its peatbogs – and without incentives not to do so it is quite possible that it will – this single action would release an incredible 155 gigatonnes, equivalent to the entire world's fossil fuel emissions over the last five years. Similarly, as clearing a hectare of rainforest releases between 500 and 900 tonnes of CO_2 – and because converting a hectare's worth of palm oil into biodiesel will save just 6 tonnes of conventional fossil fuel emissions annually – even the best regulated schemes at these locations will take around eighty years or so to offset the carbon cost of destroying the rainforest in the first place.

Bioethanol

Despite these depressing statistics, however, very much the worst alternative – yet paradoxically the one into which the US is now rushing headlong – is not biodiesel but bioethanol. This is because the industrial processes required to make bioethanol essentially take not just the land but also real food crops and turn them into fuel. It does this by fermenting the sugars and starches within what would otherwise be food in order to make alcohol. This is then refined and distilled into fuel which is suitable for powering internal combustion engines of a type similar to those that are ordinarily powered by unleaded petrol or gasoline.

This last attribute means it is well suited to the present infrastructure – the vehicles which use it also need relatively little in the way of modification to run this sort of fuel – which goes a long way to explaining its current popularity. But unfortunately making ethanol this way is extremely capital intensive, very energy intensive and very time consuming. And once it has been correctly distilled to get rid of the water content – a process which is necessary in order to allow the natural fermentation process to take place – the finished product is still around a third less efficient than unleaded petrol, meaning that vehicles using it will run for a shorter distance on a full tank.

Many countries have nevertheless been experimenting with bioethanol and ramping up production – Saab, Ford and Lotus are among those manufacturers which have recently unveiled cars that are able to run on it – while in the UK a small number of supermarkets, chiefly in the south-west, have taken steps to sell it. (Unfortunately in the UK the fuel suffers from something of a double whammy when it comes to inbuilt inefficiencies. This is because the sugar used to make the fuel is derived from sugar beet rather than cane – one of nature's least efficient sources of natural sugar.)

There are, moreover, a host of environmental problems associated with the increased use of food crops for this purpose. For example, maize mono-cropping as practised in the US is fast eroding farmland, as well as proving to be extremely water-intensive and very heavily dependent on a variety of chemical herbicides. Even more worryingly, perhaps, Uganda's President Museveni has said he plans to encourage the destruction of large areas of the Mabira Forest Reserve – his country's most important intact area of rainforest – in order to grow crops for biofuels. As much as a third of the reserve, or about 7,000 hectares of an area which has enjoyed protected status since the early 1930s, looks likely to be turned over to sugar cane production by a company based in India. With obvious implications for increased global warming, this is a highly destructive scenario, and unfortunately one which is being replicated in many other areas of the world.

Biomethane

In order to make biomethane, organic waste is allowed and encouraged to decompose anaerobically (i.e. without oxygen) so that, for example, the methane emitted naturally by a landfill site or coming from farm waste storage tanks can then be usefully employed rather than simply being allowed to leak into the atmosphere.

In theory this is a sound and workable concept, although for maximum efficiency the gas produced in this way needs to be injected into the national gas supply system as a replacement for natural or fossil gas. Unfortunately this does not happen, and instead it is typically used on site to power relatively inefficient electricity generators – inefficient because they necessarily operate on too small a scale compared with a conventional two-cycle commercial power station. From a practical standpoint the generation of fuel this way is also very slow, as the natural rotting process cannot easily be hurried or accurately controlled.

Algae and seaweed

In an attempt to supplement the above, there have been several experiments in which suitable crops are grown in totally artificial environments. One such is where species of algae are raised in vast arrays of glass tubes laid out in the desert. A network of these could in theory be made to function as a sort of bio-solar power station, the glass tubes providing a practical means of growing suitable crops in a hot, sun-rich environment while using the bare minimum of water.

Unsurprisingly the capital costs of establishing such a facility are likely to be extremely high, particularly when one includes the obvious requirement to filter and dry the algae before it can be used. Nevertheless, Algaelink, a subsidiary of the Dutch firm Bioking, unveiled what it called a photobioreactor in October 2007, pointing out that, with an oil content of up to 70 per cent and the potential to grow year-round, certain strains of micro-algae were potentially a far richer source of energy source than many other fuel crops, including soy, palm, maize and rapeseed. Such species also grow up to 100 times faster than trees, typically doubling their weight every twenty-four hours.

As we saw in the chapter on carbon sequestration the algae is at the same time capable of extracting CO_2 from the atmosphere and absorbing nitrogen from waste water. Because it can be self-contained in glass tubes, or even open ponds, such a process also has the potential to be a far less controversial source of biodiesel than, say, soybeans and other industrial crops.

Interest has also been shown, particularly in the US, in the development of so-called cellulosic ethanol. Extracted from rapidly growing mixed wild grasses, this has the advantage ecologically speaking of allowing for the reinstatement of North America's traditional prairies. As well as multiplying by a factor of sixteen the amount of greenhouse gases stored in the root systems when compared with maize, this, reported *New Scientist* in August 2007, could yield up to fifteen times the amount of energy used in its production. But unfortunately America's powerful agricultural lobby is, for the time being at least, more interested in growing maize for fuel than promoting wild grass.

Finally, consideration has been given to using sea kelp as a suitable feedstock. This too has the advantage of being an extremely fast-growing species, and as we don't yet farm the seabed, seaweeds of various kinds may well offer a possible source of feedstock for future power generation.

The carbon-neutral con

Despite considerable evidence to the contrary – and cheerfully ignoring the fact that even now the uptake of biofuels has much less to do with environmental concerns than with new business opportunities, support for farmers and national fuel security – supporters of the forms of renewable energy described above also like to claim that biofuels are carbon neutral or nearly carbon neutral.

The basis for this claim is a supposition that all the carbon which is released at the end of the cycle – that is when the fuel is being burned to generate electricity or to provide motive power for vehicles – is promptly reabsorbed and so effectively cancelled out at the beginning of the next cycle when more plants begin to grow. These in turn are harvested to make the next batch of fuel, and the cycle continues with no net increase in atmospheric CO_2, so that – their supporters maintain – no further contribution is being made to global warming.

This is nonsense, however. Biofuels are a very long way from being carbon neutral, and the cycle is by no means a closed one. Most obviously a great deal of energy is required to grow the crops in the first place, also to transport them to a processing plant and there to convert them into fuel. Fertiliser manufacture and delivery (and as observed previously monocultures of this sort are notoriously heavy users of highly energy-intensive fertiliser as well as harmful to the local ecology), the manufacture and use of the appropriate farm machinery, and of course the building and running of biofuel processing plants are similarly all considerable users of energy, and so by extension very large emitters of CO_2.

In fact, far from being carbon neutral, the amount of fuel used during the whole biofuel production-and-use cycle means that while some small savings can be made the overall gain is unlikely to be very much at all when compared with conventional fossil fuels. That was certainly the finding of the EU Joint Research Council's 2006 paper *Wells to Wheels*, which also takes into account the additional and even more damaging nitrogen oxide emissions that result from the more intensive agricultural activity required to grow the necessary crops.

The figures just don't stack up

In fact if the most positive figures for biofuels (suggesting a 60 per cent saving) apply at all it is only in the best-case scenarios, as for many other biofuels the relevant figures are much, much worse.

For example, Alex Farrell from the University of California, Berkeley calculates that maize ethanol at best releases just 25 per cent more energy than it takes to produce, and emits just 13 per cent less CO_2 than conventional unleaded petrol. Similarly, Nobel laureate Paul Crutzen of the Max Planck Institute for Chemistry in Mainz argues that if one factors in NOx emissions from fertiliser use – and molecule for molecule nitrogen is 300 times as potent as carbon when it comes to the greenhouse effect – then the benefits over fossil fuels are wiped out altogether.

And further afield the prognosis looks even worse than this, so that a 2006 study by Wetlands International was able to demonstrate that 1 tonne of biodiesel made from palm oil derived from crops grown on south-east Asia's peatlands – before clearances began one of the largest single natural carbon sinks anywhere in the world – is linked to the emission of an extraordinary 10–30 tonnes of CO_2.

Incidentally, that's without taking into account those emissions resulting from the peat-clearance fires and the loss of the peatland's inherent carbon capacity. Include these, and the report's authors estimate that a single tonne of palm oil biodiesel – an increasingly popular fuel – would be responsible for between two and eight times the carbon emissions one would otherwise have expected to get from the conventional or fossil diesel the palm oil product is meant to replace.

Water depletion is also becoming a big issue here, with the Water Management Institute, a body supported by the World Bank, reporting in September 2007 that China and India were risking famine by routing scarce water supplies into biofuel production rather than crops. As the two countries are expected to provide two-thirds of global biofuel growth in the short to medium term – India by producing sugar cane and China by growing maize – this is a risk which simply cannot be ignored.

Biofuels: a backward step

Even these daunting statistics, however, are still not sufficient to prevent supporters of biofuels from arguing that growing plants for fuel offers outstanding potential when it comes to the plant's carbon sequestration capacity. Such plants, they say, take in CO_2 from the atmosphere and, using it to grow, store it in both their own biomass and in the surrounding soil. But this is another entirely fallacious argument, since energy crops of this sort do nothing to decrease overall CO_2 emissions. It is true that an amount of the gas is captured during the growth cycle, but the same gas is then set free immediately the crop is harvested, and indeed (as we saw above) even more of it is then emitted, and at literally every stage during the production and use process.

Admittedly there may perhaps still be an argument in favour of growing these crops on waste or otherwise useless land: in this way the exclusion zone around Chernobyl could provide around 20,000 square kilometres of land on which to grow crops for uses other than food. More commonly, though, biofuel crops are grown on valuable agricultural land, on cleared forest or even on conservation lands like those in Uganda and the aforementioned peatlands. When this happens it is potentially disastrous: natural peatlands alone account for around 550 billion tonnes of carbon – that is 30 per cent of the world's terrestrial carbon – all of which is quickly released into the atmosphere if the peat is drained, cut or burned.

Clearing it for biofuels is therefore something of a double whammy for the environment – the World Land Trust, a UK-based conservation agency, is already talking of the threat of 'biodiversity losses and damage to water resources' – and certainly it looks like more of a backward step in the battle against global climate change than the advance which biofuels are increasingly said to represent. That's certainly the view of Biofuelwatch.org.uk, an independent organisation which has repeatedly expressed concern that the expansion of the biofuels industry is accelerating climate change rather than mitigating it.

According to Biofuelwatch the acceleration is coming 'through deforestation, ecosystem destruction, peat drainage, soil organic carbon losses, and the wider effects of increased nitrate fertilisation'. Moreover, reported the organisation in its April 2007 report *Biofuels threaten to Accelerate Global Warming*, the expansion is being undertaken without any proper risk analysis having been done. On the contrary: 'the wider impacts on the global climate and on ecosystems have been ignored and risks of potentially catastrophic impacts, however high or low the probabilities might be, have not been looked at'. But perhaps the final word should go to Roger Samson, a biofuels policy specialist at Resource Efficient Agricultural Production in Canada: if this is a horse race, he says, we're betting on a donkey.

Chapter 9

FOOD AS FUEL

How many Mexicans does it take to feed a Jeep?

Perhaps the biggest problem with biofuels, however, is not primarily an environmental one, and still less a technological one. Rather it's a moral one – their promotion brings with it the risk of widespread human starvation – with even the OECD recognising that with surging food prices and the destruction of natural habitats 'the current push to expand biofuel use is creating unsustainable tensions that will disrupt markets without generating significant environmental benefits'.

It has been calculated, for example, that in 2005 14 per cent of the US corn harvest was used to produce just four billion gallons of ethanol, thereby cutting America's petrol usage by just 1.7 per cent. In fact if the whole of the country's corn harvest was used to make ethanol it would still only save 12 per cent, and that's ignoring the fact that a staggering 80 per cent of the energy produced actually goes into making the ethanol in the first place.

Little wonder that such a serious strain is being put on food prices and supply, or that other observers have suggested that a 'perfect storm' of social and ecological factors is now gathering. They argue that when it breaks it will be the world's poor who will bear the brunt. The reasons, they say, are best summarised by a single, simple statistic: the grain required to fill a car's 25-gallon fuel tank with ethanol would be sufficient to feed one person for a whole year.

Why the poor's problem is your problem

For most of us in the industrialised West, this is perhaps neither here nor there as expenditure on food represents only 10–20 per cent of our total consumer spending. This gives us some leeway if and when prices rise, and anyway we enjoy a more varied diet – giving us more alternatives should the cost of a particular commodity begin to soar.

In developing countries, however, the equivalent figure is closer to two-thirds of an average family's total expenditure, meaning even a modest rise in the price of staples would be sufficient to tip many into serious hardship, possibly even starvation. This risk is also growing because almost every plant species *they* eat can now in theory be readily converted into fuel for *our* cars – including but not only cereals, maize, rice, soybeans and sugarcane – meaning the risk of price rises increases daily as the crucial dividing line between the food economy and the energy economy begins to erode.

And that's a *very* big problem, not least because while (as we have seen) there are numerous environmental difficulties associated with the use of biofuels instead of oil, gas and coal, perhaps the biggest problem is that as a planet we simply do not have sufficient productive agricultural land to allow for a significant proportion of it to be taken out of food production and replanted to grow more of these so-called energy crops.

Instead the reality is that mankind already crops more or less all the viable land globally, and even by doing this is at present only just about able to feed the current population of around six billion. (Just about, that is, because while it is true that millions starve each year this is usually a consequence of their being unable to afford the food, rather than there being an actual global shortage of food.)

We can raise yields – but not enough

That said, of course, with most crops and in many countries there is clearly some potential to increase the yield per acre, even though this can probably be done only at some financial cost and with considerable planning. There is also increasing concern about the impact greater atmospheric carbon can have on plant growth. Research from Southwestern University in Texas, for example, indicates falling protein levels in many key crops – including wheat, potatoes, rice and barley – with plants incorporating more carbon and as a consequence producing more carbohydrate and less protein.

But even if these problems were to be overcome, and were plant yields to be improved and quite dramatically, the chances are that we would still be left with the problem thanks to the additional pressures of population growth. The population of the world, after all, is rising incredibly fast. Most current estimates suggest it will grow by another 30 per cent before 2030, meaning that much more food will soon be needed, and this at a time when the likelihood is that climate change will actually reduce the amount of land suitable for growing food as well as disrupting precisely those climatic conditions we need to make the most of any new, higher-yielding crops.

Just look at a few of the forecasts contained in the 2007 *IPCC Summary for Policymakers*. These indicate drying over large parts of northern and southern Africa, most of Brazil and parts of Chile and Argentina, in Central America, across large parts of Australia, and the Middle East and Central Asia, with seasonal drying over much of south and south-east Asia. Together with temperature rises such changes cannot fail but reduce current agricultural production levels – and to do this in many of the countries where biofuels are being most heavily promoted. In Europe the signs are not good either, with oil seed rape yields not rising to meet demand but actually falling for the last three years because of what the report refers to as 'extreme weather impacts'.

Unfortunately this problem can only be exacerbated by the new and apparently insatiable demand for biofuels. As demand for these rises so will prices, meaning it becomes increasingly profitable for growing numbers of farmers – and who can blame them? – to switch out of food production altogether and into fuel crops. Thereafter whenever the value of a particular food crop drops below its potential value as fuel – and regardless of the reason – market mechanisms will automatically convert that same crop to fuel, thereby removing it from the food cycle.

As Lester Brown of the Earth Policy Institute noted, that might be a problem for some consumers in the West, what with 800 million motorists around the world crying out for fuel. But for developing countries the same scenario heralds a certain and absolute disaster: when at least two billion of the world's poorest are already having to spend half their income or more on survival rations, even a small rise can quickly and quite literally become a threat to life.

No loaves then, so how about some fishes?

Unfortunately the seas are unlikely to provide a solution. Across the world's oceans countless species are also in crisis, with volumes of fish and shellfish caught and landed diminishing rapidly, and most if not all of the world's fisheries already being exploited at wholly unsustainable levels. This is not just in terms of the take, although this is clearly causing problems to fishing fleets and fish consumers the world over, but also in terms of the sheer cost of the energy required to bring in this falling harvest.

In 2000, according to David Ehrenfeld, Professor of Biology at Rutgers University in the US, an estimated 50 billion litres of fuel – all of it fossil, and mostly diesel – was expended to land in a single year more than 80 million tonnes of fish and edible invertebrates. To put that into context it equates to an extraordinary 1.2 per cent of global oil consumption, or about the same as that used by the whole of the Netherlands for the entire year. Significantly it is also more than twelve times the energy value of the catch itself, and will have been responsible for releasing a staggering 130 million tonnes of CO_2 into the atmosphere.

Such figures naturally sound horrendous, but in fact the fleets pursuing luxury or commercially higher value species such as shrimp, tuna and swordfish are much worse, typically burning around 2,000 litres of fuel per tonne of catch. Moreover, because catches of all kinds have been declining in recent years, these same boats now have to remain at sea for longer and longer each season, so that globally the energy efficiency of the world's fishing fleets is rapidly worsening as each year passes. What this means, says Ehrenfield, is that even were we to improve the efficiency of diesel and petrol marine engines by, say, 10–20 per cent – something which given the will is entirely possible – the total amount of energy expended, and of carbon emissions, will nevertheless continue to rise in most fisheries and will do so until they finally collapse, as the North Atlantic cod fisheries did in the mid-1990s.

In one sense at least such figures and these forecasts might provide some kind of justification for increasing aquaculture or fish-farming, an industry which is already providing a substantial amount of the fish consumed in the West. But unfortunately – and even leaving aside a whole range of negative environmental impacts associated with the industry – in the context of climate change it is impossible to ignore the fact that aquaculture is itself enormously energy intensive. This is particularly so where the product is the higher-value carnivorous species, mostly because species such as salmon and trout rely on animal protein supplied by fish meal – this in turn having been manufactured using smaller, commercially less valuable fish which are caught by the very same, energy-hungry trawler fleets already described.

Clearly the oceans are not themselves going to compensate for the impending food shortage back on land.

More land producing less food

Ultimately these problems are all made much worse by the fact that far from improving as time goes on food efficiency is set to reduce as overall affluence increases.

Partly this is because richer people eat more, and indeed tend to overeat and to waste more of what they buy to eat. But it is also because, as we noted previously, virtually everyone in the world at some level aspires to live as we do. In this context that means they want to eat more meat than they do at present, and with significantly increasing affluence in the two biggest populations of all – India and China – this particular problem is already becoming acute.

Put very simply, and for reasons which are explored in the following chapter, *more* land is actually now being used to produce *less* food. Vast acreages, which for literally centuries have been used to grow subsistence crops to feed farmers and their extended families, are now being used to produce animal feed instead.

Naturally this puts a severe and increasing pressure on the food supply, a pressure made much worse when yet more acreage is set aside to grow biofuels. Food, after all, is not immune to the laws of economics: instead the price of a particular staple rises in line with the falling supply – and then rises even more as demand for that staple increases in line with population growth.

Of course at one level the existence of grain and butter mountains, and wine lakes – and for that matter the fact that in the developed world successful farmers can apparently afford to feed meat and fish supplements to herds of grazing herbivores – would seem to suggest that there is enough food for everyone, and that shortages in the world are only a matter of its skewed distribution. Until recently this has undoubtedly been true, but there is now plenty of evidence that this is changing. Evidence too that rising populations, changing dietary habits and the rise of biofuels are having a marked and almost entirely negative impact right around the globe.

We drive, they starve

Biofuels, after all, represent little more than an exchange of one man's food for another man's power-hungry lifestyle. Hence the tag 'the Great Grain Robbery' – and the fact that on the world's commodity markets prices of cereals and of oils have now more or less fallen in step with each other.

That same translation of food into fuel also neatly explains why, in February 2007, Mexico suffered what has become known as the Great Tortilla Crisis, when tens of thousands of citizens took to the streets to protest at a sudden and steep 14 per cent rise in the price of maize flour. That's nearly three times the country's rate of inflation, and was without doubt precipitated by the rise in the number of ethanol plants over the country's northern border. Gruma, the Mexican tortilla conglomerate which is part-owned by leading US ethanol maker Archer-Daniels-Midland, admitted as much: that the price hike of the Mexican staple was a result of higher corn costs.

There are many other examples of similar unrest. For example, at around the same time the brewer Heineken reported a drop in its profits when barley prices rose as a consequence of barley farmers switching *en masse* to biofuel crops. US soybean prices have similarly risen, as more soybean farmers have moved into corn. In Brazil, where sugar cane is used to make ethanol, the price of ordinary household sugar has now skyrocketed. In the summer of 2007 Italy experienced national 'pasta strikes' with the public protesting at sudden price hikes of more than 25 per cent, and most recently, because many food crops, such as corn and soybean, have long been used to feed pigs, chicken and cattle, meat prices have also begun rising strongly.

Everywhere this is leading not just to inflation – although in the UK food prices are heading for a thirty-year high in terms of annual price increases – but also to increased interest rates, making a slowdown in economic growth ever more likely. There are, for example, fears that the sort of higher food prices which risk triggering social unrest in poorer countries could also kickstart a return to the old stop-go economic cycle in the more developed ones. When wages fail to keep pace with rising living costs it's not hard to imagine employees using the occasion to rediscover the strike as weapon, and thus force a move towards higher pay.

Bursting the biofuel bubble

Little wonder then that many observers, including Alexandra Spieldoch at America's Institute for Agriculture and Trade Policy, have already concluded that, while biofuel is 'being touted by many in the international community as a new opportunity for growth for developing countries [it] will not be the panacea to resolve the global crisis around unsustainable use of fossil fuels and unsustainable consumption patterns that are supported by industrial agriculture and negatively impact the environment'.

On the contrary, says Spieldoch, 'it will not solve agricultural commodity market distortions that are associated with free trade policies and [will] have devastating impacts on other countries. And it will not serve as a magic bullet to address social needs such as employment, local ownership, and food security.'

Britain's *Economist* magazine has similarly noted that 'most experts agree: it is simply not plausible for America to gain energy security by switching from oil to corn-based ethanol. There is not enough agricultural land available. Nor is it clear that ethanol is really that much greener, given the energy needed to produce it.' Nor, it has to be said, is there enough water, The Stockholm Environment Institute having demonstrated that if 50 per cent of fossil fuel usage for electricity generation in the US switched to biofuels by 2050 it would require up to 12,000 cubic kilometres of water per year for irrigation and so on, equivalent to more than 85 per cent of the total annual flow of every river in the world.

And what in this example is true for the US is true for the rest of us too. Most obviously, if biofuel production and use expands at the rate most now expect there will simply be no choice but to reduce the amount of land currently being used to produce food or – no better – to destroy other forests, peat bogs and other areas, thereby releasing vast amounts of carbon currently safely stored away. In fact, with the population growing at its present rate the best estimates indicate that we shall need an extra 200 million hectares at least of agricultural land by 2030 just to feed everybody. Yet by this same date another 290 million hectares will have been lost to agriculture if the push towards biofuels is allowed to continue unchecked. That's in order to meet just 10 per cent of projected global energy demands, incidentally, and must be viewed against a background estimate from Sten Neilson,

Deputy Director of the International Institute for Applied Systems Analysis, of there being worldwide just 300 million hectares of suitable land which is not yet under cultivation.

This of course is why, as we saw in the previous chapter, OPEC can afford to threaten the West over the issue: ethanol and the other biofuels won't put its members out of business. OPEC's secretary-general went so far as to publicly reassure his members during an interview with a French newspaper in 2007 that they had nothing to worry about. 'As to biofuels,' he told *Le Monde*, 'OPEC knows that substitution is impossible. They will never make up more than 10 per cent of world production. This should not cause them nightmares.'

The bottom line is that biofuels are enjoying a boom, but only because they appear to be an easy option. Easy because no higher taxes or rationing are required. Easy because the delivery infrastructure mostly exists. Easy because our cars, with barely any modifications, can stay on the road this way. And easy because they keep farmers and the powerful agribusiness lobbies happy, while making it look as though our governments are actually doing something positive to address the issues which worry people everywhere.

But solving climate change, even were it possible, would hardly be this painless – and in reality, of course, most who stop to think recognise that, even if they choose, for now, to ignore it.

Chapter 10
THE IMPACT OF POPULATION

As a nation's population grows it destroys its forests – the relationship is that simple.

W hile the focus of this book has so far been on the specifics of climate change, literally every concern so far expressed within its pages is made far more urgent by the rapidly growing population of the planet – and by the fact that this growth rate continues to accelerate particularly in the most populous parts of the developing world.

Despite this, and for a very long time, it has generally been accepted that – chiefly as a consequence of increasing industrialisation, human ingenuity and much greater efficiencies – our quality of life curve will continue ever upward. Over time, this theory and past experience appear to suggest, the benefits of economic growth will cascade down to the poor in society, and – eventually – to the very poorest societies around the world. Even discounting the interrelated pressures of climate change, however, and notwithstanding the effects of increased globalisation and the continuing drive to free up world trade, the notion that a majority of the world's population will one day be able to enjoy the sort of rich and technology-rich lifestyles we in the West take for granted is beginning to look like a complete non-starter.

More does not mean better

Quite the reverse is far more likely, with a planet which is already struggling to support more than six billion individuals in all probability facing the prospect of two billion more by 2030 and another two billion in the twenty years after that. Nor is it just a question of how we feed and house all these people – although both of these pose very severe challenges of their own – but rather mankind will also have to contend with a number of different knock-on effects of rapid population growth, each one of which threatens to make what are already grave problems graver still.

Population explosions are known, for example, to make wars far more likely. A number of recent studies have shown that regardless of different racial, religious or tribal-allegiance issues within a given region one of the key predictors of armed conflict is invariably the proportion of males aged between fifteen and twenty-nine. Unfortunately, and almost without exception, the fastest growing countries of the world tend to be characterised by large numbers of restless, unemployed, unmarried and disenfranchised young men, a point well made by Gunnar Heinsohn at the University of Bremen who compared peaceable Tunisia with war-torn Afghanistan – two relatively poor Muslim countries, one of which has a birth rate nearly four times higher than the other.

Rapid population growth also has serious (if unsurprising) implications for the environment. For example, the American ecologist Joseph Wright has been able to demonstrate that a very strong inverse relationship exists between human population density and forest cover. As a nation's population grows it destroy its forests – the relationship is that simple – thereby threatening not just plant and animal species but potentially our own as well, by providing yet another enormous source of damaging atmospheric CO_2. Recent estimates such as that of William Laurance of the Smithsonian Tropical Research Institute suggest that up to 5 billion tonnes of the gas are released into the atmosphere each year by burning forests in this way – to which one must add the not inconsiderable emissions resulting from the normal, everyday lives of the millions who then crowd onto the land to cultivate and inhabit the space formerly occupied by that same forest.

Shock and ore: the realities of resource depletion

But actually, in the context of population and climate change, the smoke from burning trees, the loss of forested areas' natural ability to absorb vast quantities of CO_2 and the even greater emissions of a larger population, are only the beginning. After all, these much larger populations also consume much larger quantities of natural resources. That doesn't just pose a problem of supply but also throws an important spanner into the works for anyone clinging to the idea that science and new technology will provide a solution to the problems we face as a result of rising temperatures and changing weather patterns.

Here, and most significantly, it's not simply the sheer number of people that causes the problem. There's also an important multiplier in that more of them wish to emulate our lifestyles and can now afford to do so. Rapidly increasing personal wealth – particularly among the vast emerging middle classes in China and India – means that potentially hundreds of millions will soon be able to afford to buy and consume at or approaching Western levels. The effect of this multiplier on fossil fuel resources hardly needs explaining, but the fact that more people are now buying more hi-tech goods is putting an immense, unsustainable strain on many increasingly key natural resources.

Such a strain, that is, that it has already been noted (by observers such as Tom Graedel writing in the *Proceedings of the National Academy of Science*) that, even looking at existing rather than future technologies, 'stocks of several metals appear inadequate to sustain the modern developed-world quality of life for all of Earth's people'.

Take the case of platinum. While commonly associated with expensive modern jewellery, it also has a number of highly significant industrial applications. It is, for instance, a crucial part of every one of the many millions of catalytic converters now fitted to cars, trucks and vans in the West. Perhaps even more significantly it is an important component in at least one of the transport sector's long-hoped-for silver-bullet solutions, namely the new, clean and renewable hydrogen fuel cell – or rather it could be if we have not already exhausted supplies of the metal by the time this particular technology is perfected and put into production.

Throwing away the future

The chances are that we will have, however. We are currently consuming and trashing platinum and many other precious metals at such an alarming rate that supplies are already considered to be in jeopardy. All those of us driving catalyst-equipped vehicles, for example, which also use vast quantities of rare rhodium and palladium, every day dump literally tonnes of valuable materials onto the streets through our exhausts. So much of it, say researchers at the University of Birmingham, that in the UK an estimated 1.5 parts per million of roadside dust is now likely to be pure platinum.

So far, at least, there is no known mechanism or process for either the recovery or recycling of this precious metal waste, let alone a system in place to do this. Yet in a pattern we see repeated many times, catalytic converters are still held up as an environmental success story rather than what they are – which is another industrial-scale means to squander what was from the start a very precious, decidedly finite resource.

In fact there is so little platinum left now that, as *New Scientist* observed in May 2007, were the world's approximately 500 million road vehicles to be replaced with a similar number powered by those aforementioned and effectively non-polluting hydrogen fuel cells – something which from a climate-change perspective might be highly desirable – there would be no platinum left anywhere in the world within just fifteen years.

Consider too that unlike oil – or for that matter diamonds, another much sought-after substance with both lifestyle and industrial applications – when it comes to replacing many materials such as platinum there is simply no viable synthetic alternative. Nor, because it is a naturally occurring chemical element, one of nature's fundamental building blocks, is there any way we can create or refine more of it once the existing supplies have been used up.

Technology more of a problem than an answer

In this way it can be seen that, far from solving our climate-related problems, many of the most vaunted new technologies are far more likely to fall victim to supply problems or, if not, to severely exacerbate these if and when they finally come on stream. This is because, all too often, we use the gains from new energy technologies simply to continue to increase our consumption and waste. Eventually that puts us into a vicious spiral, which far from improving the situation serves only to decrease resources and increase environmental damage – and to enable us to continue doing this even as our energy technology improves.

Perhaps the best example of this particular difficulty is one of the so-called rare earth metals – indium – supplies of which will almost certainly be exhausted before the man or woman on the street has even heard of it. A couple of decades ago the stuff was virtually unknown and more or less worthless, and in 2003 it was still only trading at around $60 a kilogram. But at the time of writing that situation was changing dramatically, with *New Scientist* not alone in reporting its 'impending scarcity' and indicating that this was now reflected in its market value, the price having shot up to 'over $1,000 per kilogram' in less than three years.

The rocketing demand and rocketing price are both easily explained: indium had become an important component in the manufacture of LCDs for all those increasingly popular (and now worryingly affordable) flat-screen televisions, computer monitors and big-screen mobile telephones. Of course it could be argued that, if push comes to shove, we can always make do without our flat-screen TVs. But unfortunately, and together with another metal called gallium, indium is also used to make a novel semiconductor material, one which turns out to be crucial to the efficient functioning of a brand new type of solar cell.

According to its creators, this cell promises to be around twice as productive as conventional photovoltaic technology. The trouble is, and long before it is anywhere near being made available commercially, world supplies of both indium and gallium are running out. Worse still, they are doing so at such a rate that in 2007 a chemist at Leiden University calculated that reserves of both were *already* insufficient ever to allow the still-to-be-launched solar cell technology to make what he called a 'substantial contribution' to the future generation of clean electricity.

In fact, reserves of indium are likely to be exhausted in as little as ten years, meaning that at least one possible solution to the search for clean, sustainable power is already more or less dead in the water. And unfortunately there are plenty of similar examples for anyone who cares to look. Tantalum is needed to make a variety of compact personal electronic devices, such as mobile telephones; hafnium is a vital component in the manufacture of silicon chips; and germanium is used in a wide variety of infrared optics throughout the semiconductor industry.

All three have never been anything but very rare – by comparison gold and even platinum are common. While it is true that precise figures for their reserves in North America, Russia and China are not widely publicised – such information is invariably regarded as commercially highly sensitive – all the signs point to their being totally inadequate to meet demand for more than a few years.

Demand for them at current levels, that is, although the reality is that this can only rocket upwards as developing nations seek to follow the lead of the developed ones, as their citizens seek to share the same technology-dependent lifestyles which we have enjoyed for years – and as we ourselves urgently seek to devise yet more new technologies to combat climate change and replace our existing dependency on fossil fuels.

LCD versus DRC: how demand leads to wars

Even put like that it may not sound like a major issue – our lives do not literally depend on a limitless supply of new mobile telephones and flat-screen TVs – but actually the implications are far more severe, and we are already seeing the results.

Armed conflicts, for example, can and do occur when resources run short, and on occasion the supplies in question can be quite arcane. The fact that the vast majority of people have never heard of tantalum did not prevent a bloody civil war being fought over supplies of it in the Democratic Republic of the Congo. Lasting from 1998 to 2002, these hostilities coincided perfectly with a sudden surge in the price of this rare earth metal – the DRC being home to Africa's richest supply – making it just one of the more tragic side-effects of the soaring popularity of mobile communications equipment around the world.

For this reason alone, it seems sensible to ask – as does Armin Reller of the University of Augsburg – what might be the result if another resource-rich country such as China decided to turn off the tap, stopping exports of rare earth metals in order to use them in its own rapidly developing, increasingly high-tech industries? The question needs to be asked because, according to the US Geological Survey, America currently imports 90 per cent of its rare earth metals from China. At the same time the Chinese are well aware of the rising value of these materials, and are busy supplementing their own natural deposits by investing heavily in Africa and its mines, and by bulk-handling the West's high-tech waste in order to salvage any precious materials they contain, before discarding the rest.

It's not just the rare stuff either

In fact, this particular problem goes way beyond the sort of obscure materials which most of us have never heard of. Quite common ones are under threat too, such as copper, which according to *New Scientist* is set to run out before the end of the century if last year's per capita consumption figures are taken as a guide. The magazine observed that lead and tin could run out even sooner, while supplies of silver, a precious metal used in a vast array of different industrial processes, could be exhausted in as

little as fifteen to twenty years, prompting fears that the
price could rocket from $13 an ounce to $130 over the
next eight years.

That fifteen to twenty year figure assumes that
predicted new technologies appear as planned, and that the
population continues to grow at its present rate. What it
doesn't take account of, however, is the aforementioned
multiplier, a significant omission, because while it is true that
around 300 million Americans currently lead the world in the
consumption of almost everything that is non-renewable – and
their lead is very considerable – the real threat to resource
depletion will come as more than six billion non-Americans
begin to play catch-up.

That was certainly the message of the GEO-4 Global
Economic Outlook published by the UN Environment
Programme in October 2007, a fully peer-reviewed study by 400
experts which concluded that we need 1.4 Earths simply to sustain
the average lifestyle of today's six-billion+ inhabitants. If we don't
slow down, the report said, 'humanity is at risk'.

Similarly, in what it called its Earth Audit a few
months earlier, *New Scientist* had calculated that resources
which might be expected to last, say, fifty years given current
global consumption will more likely be exhausted in fewer
than twenty if this multiplier is taken into account. To
support this they provided dozens of examples, showing the
likely impact as increased affluence allows many more non-
westerners to adopt elements of western lifestyles.

Global reserves of phosphorus, an essential ingredient
in so many agricultural fertilisers, drop from 345 years' supply
at current demand to less than 150 if the rest of the world
consumes it at just half the rate of US farmers. Uranium
similarly drops from fifty-nine years to just seventeen, posing
all sorts of problems for those who propose nuclear energy as a
greener if somewhat controversial solution to concerns about
global warming. And pharmaceutical antimony, another

substance which barely measures on the public awareness radar but which we nevertheless all depend on, could be gone in as little as thirteen.

Food for all?

Yet even so the truth is that nobody needs to root around for research data on exotic-sounding obscurities such as antimony and germanium in order to put flesh on the bones of this particular problem. Instead you just need to look at another substance, one which is quite ubiquitous and even more essential, namely the food that we eat. And once again it's not just that more people require more food, but also that more people are starting to eat more like us – as mentioned in the previous chapter.

At its most basic this means they are demanding and consuming more meat and other proteins. Imports of cheese into China, for example, have more than doubled in the last five years, thereby helping to push up prices as far away as UK supermarkets. Looked at in purely calorific terms – that is, viewing food simply as energy – it also makes it clear that we face yet another immense challenge to the world's natural resources, and do so even without taking into account the previously discussed threat posed by more crops being turned into biofuels in a bid to mitigate the threat of climate change.

Most obviously more people eating more meat poses a problem, because even with increased yields per animal meat production not only generates an awful lot of climate-warming pollution but also consumes a huge amount of valuable natural resources. In fact it has been calculated by the National Institute of Livestock and Grassland Science in Japan that the creation of just a single kilogram of red meat is responsible for generating more greenhouse gases and other pollutants 'than driving for three hours while leaving all the lights on back home'. At the same time, viewed as economic units of production, farm animals are woefully inefficient when it comes to their core function of converting their food into ours.

The inefficiency of animals

Typically the conversion rate for turning plant matter into meat is no better than 10 per cent, meaning that in very round numbers it takes approximately 10,000 lbs of cattle food to make a 1,000 lb cow. Or as Jared Diamond put it rather more graphically in his Pulitzer Prize-winning book, *Guns, Germs and Steel: The Fates of Human Societies*, 10 lb of pasture, cereals and so on to produce just 1 lb of steak, or four quarter-pounders. Meat from chicken is admittedly somewhat better – requiring perhaps just three times its weight in cereal – but even so this goes a long way to explain why barely a dozen of the planet's nearly 150 large terrestrial herbivore species have ever been successfully domesticated, and also why (with only very few regional exceptions) we don't eat land-based carnivores.

It's not so much that many of the latter are dangerous and don't taste very good – although both of these statements are perfectly true. Rather it is that before being ready to eat a 1,000 lb carnivore would have needed to consume 10,000 lbs of herbivore, which in turn would have needed to consume a mammoth 100,000 lbs of vegetation. And whichever way you look at it that's an awful lot of agricultural land just to provide you and me with a mere one-hundredth of that weight in, say, lion chops or bear burgers – so we eat the herbivores instead.

Unfortunately, though, even doing it this way the aforementioned 10 per cent efficiency ratio still means that a lot of the most productive agricultural land – land which could potentially feed many millions of vegetarians – ends up being used to grow animal food in order to satisfy the demands of a much, much smaller number of omnivores. The changing diet of tens of millions of newly rich Indians and

Chinese – who for the first time in history are beginning to mimic western eating habits as well as lifestyles – makes this particular problem more acute than ever. It puts another huge strain on world agricultural production, and would do so even without the impact of climate change and the ongoing race to grow biofuels.

Intensive agriculture is not the answer

Of course one obvious answer would be to boost agricultural yields, and certainly the technology to do this exists. In fact there is plenty more of it in the pipeline, although it needs to be recognised that greater use of artificial fertilisers will increase considerably existing levels of nitrous oxide emissions.

Moreover, and even leaving this aside, the reality is that in the West at least opinion is turning against the more intensive farming methods which for years have dominated farming in Europe, Australia and North America. Mostly this is because, while it is true that many of the methods employed are more efficient, at least in terms of yields per acre, chemical fertilisers and pesticides have fallen out of favour with consumers who feel they are unhealthy and unnatural.

In this context there is a sustainability issue too: the use of such products also helps to make modern agriculture a massive consumer of energy and of scarce resources while also damaging local ecologies and as a consequence reducing biological diversity. But unfortunately the most popular alternatives, low-input or organic production, simply cannot match the high yields of conventional, hi-tech farming.

In fact in Britain while the average organic wheat yield per hectare is around 4 tonnes the average yield on a conventional farm, says Edinburgh University's Anthony Trewavas, is closer to 8 tonnes. In other words we would need twice as much land to produce the same amount of food. Of course the first problem with this equation is that we don't need the same amount of food, we need much more of it (what with the world's population set to grow by 30 per cent before 2030, giving us another two billion mouths to feed). And the second is the near-certainty that in the future we will actually have less land to farm rather than more – not just because a growing population means more land will be built over but more significantly because changing weather patterns and the very real threat of rising sea levels will soon take much of it out of production altogether.

Floods, droughts and feeding the multitudes

You don't even need to leave the rich and agriculturally sophisticated developed world to see this. Most experts acknowledge, for example, that at a time when we are becoming much more accustomed to strange or freak weather conditions it would take just one decent storm in the North Sea coinciding with a spring or high tide to inundate up to 300,000 hectares of Grade 1 English agricultural land with seawater.

Those valuable 300,000 hectares occupy the area we know as the Fens, some of the richest but lowest-lying agricultural land in the whole of Europe, and of course it is true that were such a set of circumstances to occur the floodwaters would eventually subside. It would, however, take literally years to desalinate and decontaminate the soil after such a flood, and be many more years before anyone in the region would see a return to its current, highly impressive levels of agricultural productivity.

Then there's the example of Australia, where the so-called Big Dry – while associated with El Nino but by no means a mere blip or simple climatic anomaly – has turned into a really major drought with massive implications. At the time of writing – a full six years after it started, and with 2005 the hottest year on record – it shows little sign of abating, making it hard to take comfort from the fact that Australia is fundamentally a dry country, has always been a dry country and has experienced plenty of droughts before.

All of that's true of course, but it has never been quite like this. In fact the Big Dry is thought to be the country's most serious drought for a thousand years. It's certainly the worst in more than a century of recorded data, with rainfall totals falling consistently for the last five decades and climate experts forecasting a further drop over the next fifty years.

Because most of Australia's more than two-dozen coal-fired power-stations rely on huge quantities of water to generate their power, the risk of powercuts has risen sharply — as has the cost of electricity, which industry observers say could rocket by up to 30 per cent in a single year. Wildlife has suffered badly too so that,by mid-2007, the kangaroo population was reckoned to have halved, while in New South Wales sharks had been seen 25 miles inland having swum upstream in places where river levels had dropped so far that saltwater was able to extend further inland than ever before.

Local problem, global impact

But of course the real casualty of the Big Dry is food production, with a full 70 per cent of the country badly affected and the worst hit area being the Murray-Darling river basin. Besides being home to more than two million Australians, this area has long been Australia's food bowl with fully three-quarters of the country's entire stock of irrigated agricultural land accounting for around 40 per cent of the country's total food output. By 2007, however, that position had changed dramatically with the wheat crop down by 61 per cent, barley by 63 per cent and rapeseed or canola cut by an incredible 71 per cent. Rice and wine growing has fared even less well and collapsed more or less completely, while tens of thousands of emaciated sheep and cattle have been sold and slaughtered at wholly unprofitable prices.

It's a long way off from Europe, but the lessons from this could scarcely be clearer: despite considerable ingenuity over the last fifty or more years, dramatically increasing crop yields in Australia and elsewhere using a variety of biological and technological advances, agriculture and by extension all of us remain very much at the mercy of the simplest climatic factors.

Rainfall, sunshine and temperature – it really is that simple

Even leaving aside such a dramatic example as the Big Dry, the reality is that we need to consider that it would take only a relatively small change in average temperature to produce such a scenario closer to home. Just a degree or two, that is, to markedly alter forever the agricultural profile of virtually every country on the planet.

Were this to happen in the northern hemisphere, for example, it would as previously noted be necessary to shift agricultural production to cooler climate zones futher north. In fact, according to the journal *Nature*, between 38 and 52 per cent of all species, flora and fauna, will need to relocate to other geographic ranges in order to survive the published estimates of climate change between now and 2050. In one sense this move has already started, and several big French wine growers have been acquiring land in southern England, a region where the quality of wine is fast improving. North American wheat production might similarly shift from the mid-west to Canada since, with two-thirds of global water consumption already being accounted for by agriculture, farmers there – as in Australia – are now seeing that they cannot simply irrigate their way out of the problem.

Looked at in the round, of course, few things are ever this straightforward and indeed many of these new climate areas will quickly prove to be wholly unsuitable for large-scale agricultural activity of any sort. Areas which were recently glacial, for example, may indeed inherit the right temperature profiles over time, but they will still be characterised if not by bare, scoured rock and scree then by thin, new soils completely lacking in the necessary levels of nutrients.

These will be nothing like the rich soils we actually need, and in the absence of the natural resources required to boost their fertility artificially it would take many centuries for them to develop and mature to a point where the areas in question can come even close to matching the fertility of the more traditional agricultural areas of Europe, North America and Asia. Instead, as temperatures and sea levels rise, the likelihood is that agricultural yields will fall in almost every country of the world, and that both subsistence and commercial agriculture will suffer very badly as a result.

Chapter 11

MIGRATION AND MELTDOWN

More than a third of the world's population lives in areas where there is already insufficient water to meet the most basic needs.

Civilisation can never be taken for granted, which is why many organisations and individuals – official, academic and amateur – have for years engaged in the intellectual exercise of assessing how different disaster scenarios might play out in the future. It is clear from their various deliberations that, when looking for potential sources of catastrophe, we have plenty to choose from.

Recently much of the emphasis has been on an array of different 9/11-style attacks on key cities, perhaps using a low-yield nuclear or so-called 'dirty' bomb. Consideration has also been given to the possibility of a missile attack from one of the more recent recruits to the nuclear family, in the Middle East or south-east Asia. Economic rather than military assaults have also come under the spotlight: for example, the possibility that a more militant Saudi Arabia, its Wahhabi tendency tiring at last of the West's imperialist meddling and might choose to cut off the flow of oil or divert it to China and India. Russia might act in a similar fashion with its own rich gas reserves, as might China itself, newly industrialised and with its own mineral resources to draw on, deciding to flex a little muscle on the international stage. Finally there is a battery of possible natural disasters to consider as well: a bird flu pandemic, perhaps, the long promised rupture of the San Andreas fault, or a devastating comet strike.

The truth, though, is that we don't really need anything this dramatic to precipitate catastrophe, and increasingly researchers are coming to realise that all it would take to trigger some kind of social meltdown is a relatively small rise in average temperatures coupled with a few dollars on the price of a litre of petrol. Certainly the UK Government's *Stern Report* recognised this in 2006, stating quite explicitly that climate change could bring about economic and social disruption on the scale of the 1930s depression and two world wars. The result of a detailed, sixteen-month climate study, it forecast not just significant declines in world food production but also severe water shortages affecting as many as 4 billion people and a rapid fall in living standards around the world.

Even so, the likelihood is that society will unravel slowly to begin with, a so-called slow crash rather than a sudden catastrophe of the sort more usually envisaged, one which makes itself felt in a gradual dislocation from normality as average temperatures, sea levels and energy prices all begin to creep up.

How climate wrecks communities

Some early manifestations of this can be seen around the world already. Not just when we look at incidents involving increasingly freakish weather patterns in many different climate areas, but also in the very significant cuts in agricultural yields we are witnessing – sometimes because of drought, at other times as a consequence of severe or prolonged flooding – and the economic uncertainty which flows from this. Such occurrences make it abundantly clear that when climate change really begins to bite some countries will be affected far more than others. They will also be affected far sooner than others, and in different ways, although the certainty is that no country will escape unscathed.

Generally speaking, and for obvious reasons, the very poorest and the landless suffer first and most badly. It is true that some authorities, such as America's Environmental Protection Agency, have suggested that even a doubling of atmospheric CO_2 would lead to only a very modest drop in global crop production. But the reality is that more marginal communities cannot afford even a tiny drop in overall yields, so they feel the effects very early on.

While there were of course other factors besides climate change, the World Meteorological Organisation estimates that between 2001 and 2003 there were a staggering 854 million undernourished people across the globe. Unsurprisingly more than 96 per cent of these were living in developing countries, precisely those areas of the world where, according to the Organisation's Secretary-General M. Michel Jarraud, 'subsistence farmers depend on rain-fed agriculture for their survival and the productivity of crops on their farmlands is impacted heavily by seasonal to inter-annual climate variability'.

Nor can one ignore the fact that it is these developing countries in lower latitudes which look set to suffer disproportionately at the hands of climate change. It may be true, as the WMO asserts, that great strides continue to be made in agricultural productivity around the world, with yields improving year on year. But mostly this is not in these regions, where the tendency is still negative: towards more extreme climatic conditions such as floods, droughts, forest fires and tropical cyclones, and with predictable results for soil degradation, overall productivity and the food security of their inhabitants.

Food and water: nothing else matters

Human life, after all, is essentially about food and water. Everything else is secondary, and increasingly – particularly in areas where populations start to swell as a result of inward migration – everybody's problems will come down to the same thing. That is, the availability of food and water and the means by which communities secure supplies of each as demand for both rises and the availability begins to reduce.

In fact we are already seeing precisely this scenario. The civil war in Darfur is fundamentally about Arab and African competing over scarce water resources. Nor are these sorts of climate-related conflicts by any means a new phenomenon – the journal *Human Ecology* recently reported the findings of an historical survey by the University of Hong Kong indicating just such a link in twelve out of fifteen major Chinese wars – but clearly it is one which is going to become far more prevalent as the effects of climate change continue to make themselves felt.

For example, only 1 per cent of the world's water is fresh rather than saline, yet our consumption of it has rocketed by more than 1,000 per cent over the last 100 years. It's true that desalination is possible, but it is also hugely expensive in terms of both capital outlay and energy, and anyway water is still subject to a finite supply. Because of this, Koichiro Matsuura, Unesco Director General, reported as long ago as 2003 that 'over the next twenty years the average supply of water worldwide per person is expected to drop by a third'. More recently it has been calculated that an incredible 2.3 billion people – more than a third of the world's population – already live in areas where there is insufficient water available to meet the most basic needs of drinking, food production and sanitation. Also that 1.7 billion of them are surviving in conditions of true water scarcity, with less than 1,000 cubic metres per person per year.

The situation regarding food is scarcely any better. In 1986 the World Bank defined food security as 'access by all people at all times to enough food for an active, healthy life' — on which basis we can see that even now we are failing to achieve full food security. (This despite the call at the World Food Summit more than twenty years ago for concerted action among all countries to achieve this goal in a sustainable manner.)

As long ago as 1980 the World Food Trade Model put the number of people at risk from starvation at 500 million. The figure excluded China, where figures were hard to come by, but in any event it is now thought that the numbers will climb to 640 million by 2060. That's without allowing for the effects of climate change, which dramatically worsen the situation, in particular because crop yields are set to decline even more in lower latitudes. Not unnaturally this brings with it the threat of much higher food prices, and unfortunately higher prices necessarily increase the number at risk from starvation, typically by around 1 per cent for every 2 per cent increase in food prices. This obviously provides yet another impetus for more environmental migration, and of course for social unrest.

Unfortunately climate change disrupts communities in other ways too, as was shown when, in a bid to explain the high number of civil wars in sub-Saharan Africa, economists Edward Miguel, Shanker Satyanath and Ernest Sergenti examined rainfall totals in the region. They found that an incredible twenty-nine out of forty-three states were involved in some kind of civil war in the 1980s and '90s, and that while the causes were often complex, low rainfall (and the resulting agricultural and economic shocks) was among the most reliable predictors of trouble. What they described as a '5-percentage-point negative growth shock' was shown to 'increase the likelihood of civil war the following year by nearly one-half'.

Still with water, the potential for more Darfur-style conflicts is further heightened by the fact that a full 40 per cent of the world's population lives within river basins which are shared by two or more countries. As one is invariably better off (in water terms) than the other, and as drought leads inexorably to food shortages which can quickly breed resentment, it is not at all uncommon for the effects of climate to translate into popular uprisings, the destabilisation of authority and even actual invasion.

A rising tide of environmental migrants

Inevitably, with so many communities in danger of being rendered non-viable in this way, literally millions of people are now being forced to flee their homes and homelands. The result has been to create a new class of what one might call environmental migrants as these people seek refuge in more secure, stable and prosperous areas. Now, increasingly, climate change will force both internal and cross-border migrations as people leave areas where food and water have become scarce, and as they flee rising seas and areas devastated by the droughts, severe storms and flooding that are the expected consequences of rising temperatures. Clearly South Asia, Africa and Europe will be particularly vulnerable to these mass migrations.

At present, however, such individuals do not belong to any recognised class of refugee, and as such they have no official status or protection under the Geneva Convention. They are certainly with us though and indeed their numbers are already thought to exceed those of conventional or political refugees, and by several million. As early as 2003 the total was put at 25 million, although with a certain tragic inevitability even this level of personal tragedy has still not yet led to any serious initiatives being enacted on their behalf.

This lack of any meaningful action is to a great degree a function of definition, or rather the lack thereof. If they are classified at all, the displaced are likely to be thought of simply as economic refugees, a convenient categorisation which manages somehow to convey the impression that they are seeking voluntarily to better their chances with an easier or more comfortable life in a more stable society. In this way richer communities can turn them away without experiencing too much in the way of guilt. Even a cursory examination of the problem, however, demonstrates that these are people who genuinely have no choice but to relocate or die. It also makes obvious the fact that sooner or later the problem of environmental migration will have a major impact on the developed world as well.

Millions on the move
This is particularly true as the aforementioned figure of 25 million is certain to be dwarfed once climate change really gets a grip: Andrew Simms, Policy Director at the New Economics Foundation (or NEF), calculates that by 2050 the number could be in excess of 150 million. Even this figure sounds conservative when one considers the 165 million people affected by climate-related disasters in 2001 alone, the more than 100 million who over the previous decade have fallen victim to drought and famine in Africa, and indeed the possibly even larger numbers who were similarly affected by flooding in Asia over the same period and again more recently.

For the developed world such numbers sound a warning signal that we too are heading for trouble, most obviously because what we have seen so far is only the fallout from desertification and heavy rainfall. From now on, once sea levels start to rise – and with it the numbers of environmental migrants – these problems will be of a completely different magnitude.

Aware of this real and imminent threat, at least one of the smaller and more vulnerable nation-states has taken matters into its own hands already. Low-lying Tuvalu in the South Pacific has, for example, signed an agreement to allow phased evacuation of its citizens to New Zealand when the time comes. Of course, with an area of just 26 square kilometres, the total population of Tuvalu – where increasing salination of its soil is already causing problems to agriculture, and the landmass is itself shrinking – is a mere 12,000, making such a move relatively easy. Even the Maldives, its 1,196 coral islands and atolls totalling 298 square kilometres, is home to only 350,000, so a similar programme can be enacted here before its peak elevation of just 2.4 metres allows it to sink beneath the waves.

But clearly far more serious problems face those countries with longer coastlines and much bigger populations – especially when one considers that the lowest-lying 2 per cent of the world's total land mass is home to more than 10 per cent of the world's entire population. NASA climate expert James Hansen has shown that without really radical changes to the way we obtain and consume energy the resulting rise in sea levels will be sufficient to flood the homes of more than 50 million Americans, 250 million people in China, particularly around the mouths of the Yangtse and Yellow rivers, 150 million in India, and virtually the entire population of 120 million Bangladeshis.

Clearly in the face of such a danger these people too will attempt to relocate rather than die – but where will they go and how will they survive? What will happen to the more than 12 million who could be displaced in the Philippines, the millions more in Cambodia and Thailand, and the four-fifths of Egypt's 80 million population who live in the highly flood-prone region around the Nile delta?

No masterplan, and no-one doing any planning

The truth is, nobody knows. Certainly no-one in authority has yet taken steps to properly evaluate the problem, to construct a rescue plan or to establish how these mass migrations will have an impact on the rest of the world. Even the logical authority to carry this out, the UN High Commission for Refugees or UNHCR, admits that it simply lacks the necessary resources to tackle a problem of this size. Instead, it says, the truly vast scale of the expected migrations means they will have to be organised and controlled at national level – although, as Andrew Simms of the NEF was quick to point out, with something like 20 million new environmental migrants on the move each year, 'the national level may be under water'.

All we have seen so far is the predictable call to create new legal obligations, chiefly on the part of the richer nations to accept these environmental refugees. Thus NEF proposes that as the UK is responsible for 3 per cent of global emissions then morally it should take in at least 3 per cent of these displaced persons every year. There is a certain logic to this, not least because those countries with only marginal responsibility for global warming otherwise look set to pay the highest price for it, while the rest of us carry on enjoying the freedom to consume and pollute, and to do so without pausing to consider that this freedom might actually have a price tag.

But even the staff of NEF must know that nothing of the sort is going to happen, and that such huge numbers of refugees will be far from welcome. For one thing it would be fiendishly difficult to arrange: 600,000, the number allocated to the UK in this way, is equivalent to all the Eastern

Europeans who have arrived in this country since 2004. It is also well over four times the number of conventional migrants accepted into Britain annually according to the last set of official figures (139,260 in 2004). Such a move would probably be politically unacceptable too, and this even in a relatively wealthy, relatively well-regulated country such as our own, one with an historically good if occasionally fragile record when it comes to taking in the stressed and oppressed.

The West destabilised

Clearly then we are heading for trouble. It may be true, as a British prime minister once famously remarked, that the places currently feeling the worst of it are 'far away countries . . . of whom we know little'. But the reality is that even if we and other developed countries choose to reject the NEF's encouragement to do the decent thing, and take no action whatsoever, the refugees will be heading here anyway. Increasingly they will have no choice but to leave home and to do so *en masse*, and it is inconceivable that Europe and the other developed economies will be anything but the destination of choice for those who can make it this far.

Once that happens, and regardless of how well or badly the migrations are managed, the impact cannot fail to be enormous. The movement of many millions of poor and dispossessed across national boundaries will inevitably cause massive resentment and bitterness, and not just among those who are expected to take in these huge numbers of outsiders (and to feed, clothe and house them) but also among the refugees themselves, many of whom will not wish to be second- or third-class citizens in an alien and relatively unwelcoming environment.

The result is a potent cocktail and one which, even in normal times, would bring with it the capacity to threaten the stability of even the most secure and well-organised society. The reality is, though, that by this time things will be far from

normal. Rich countries such as our own will also be
undergoing the effects of climate change, and experiencing
disruptions of our own as food and fuel prices rise and the
pressures on our own society begin to multiply.

To see that it's not just the developing world which is
affected by these things we need to look again at Australia's
Big Dry. It is now widely accepted that in part at least this is a
result of more frequently occurring and more intense El Nino
activity, itself brought on by climate change. It remains true,
however, that to a very large degree the Australians have
themselves magnified the problems considerably and continue
to do so.

It is true, for example, that by burning so much coal
Australians have for some while had the highest *per capita*
greenhouse gas emissions in the world after Luxembourg. It's
also true that for more than a decade they continued voting for
a prime minister who admits to being a climate change sceptic
and for a succession of governments which have dismissed the
Kyoto protocols out of hand. (Those governments also
permitted unrestricted access to cheap water for irrigation and
for tens of thousands of private swimming pools.) And it is
equally true that the country has aggressively pursued an
agricultural policy based around two completely alien species,
both of which are notoriously water-intensive – namely cotton
and rice – and that the last of these is also a very significant
producer of methane, another major greenhouse gas.

Everybody feels the impact
But no matter. Globalisation means that Australia's problems
are now ours, thanks to the sort of mechanisms which these
days ensure that – as this sort of climate-driven scenario plays
out elsewhere – everybody gets to feel the effects. For example,
because Australia has historically produced around 15 per cent
of the world's wheat, the Big Dry has helped to push the price
to a ten-year high right around the world. At the time of

writing British shoppers were already paying more for a loaf of bread than they did a year ago, and unfortunately – because farm animals eat grain too – the knock-on effect of cereal prices doubling in less than a year has been to see the price of milk and cheese rocket as well. The price of even basic Cheddar has climbed by more than 250 per cent over the same period.

There are other factors too, of course. We saw in Chapter 9 that this particular problem can only worsen as the switch to biofuels continues to accelerate. The result is the same, however: an ever larger proportion of the population comes under pressure financially as the spiralling price of basic commodities affects a whole range of food prices, therefore pushing up inflation. This in turn raises the spectre of higher interest rates, which in themselves increase the threat of further inflation; together with rising fuel prices, this could cause a real and substantial slowdown in global economic growth.

Western wealth is no defence

It is of course true that, because we are richer, we have more space to manoeuvre. Because we are richer we have more slack in the system so, for example, modest price rises and reasonable shortages do not cause most us major problems as their effects can to a degree be mitigated. But none of this will hold true for ever, not least because the present model of a typical western economy works on growth rather than stability, and that in turn relies on cheap energy. What this means is that even were oil extraction to plateau rather than to disappear – and that's what's happening now, although it is sometimes euphemistically described as being due to 'insufficient refining capacity' – the very real risk is that that our whole economic house of cards could collapse, and quickly too. The very strength of the developed world paradoxically makes us the most oil-dependent countries of all.

In the face of this the risk of collapse is made worse because we are already seeing a number of structural weaknesses emerging in countries such as Britain. The traditional social fabric has been unravelling here for years, and to a point where without the sort of social cohesion which characterised the country sixty or seventy years ago it is hard to imagine any sort of community or 'Blitz' spirit riding to the rescue in order to see us through what promises to be a very painful period of social transition.

Aside from any other considerations, it is now known that high rates of immigration – while economically almost certainly an advantage to the host nation – can, as a consequence of the greater ethnic diversity which results, act to reduce the levels of trust and social cohesion which might otherwise pertain. In an item headed 'Mistrust rises with social diversity' in September 2007, *New Scientist* reported research at Harvard University involving more than 30,000 American residents and demonstrating that inhabitants of more diverse communities 'tend to withdraw from collective life, and to distrust their neighbours, regardless of their skin colour'. Interestingly the same individuals tended to have 'lower expectations of government and community leaders . . . in short, diversity seemed to erode trust and community cohesion'.

Societies fracture when the heat's applied

When the pressure is on, a breakdown of society is a very real possibility. People in the West may not know it, but for decades we have been living in a time for plenty. Cheap and sufficient food and water, an abundance of energy. But when people cannot travel to work because oil scarcity has made transport difficult, businesses grind to a halt. When this happens people don't earn any money, and their worries naturally turn to concerns about food and water, concerns which against a background of climate change begin to look all too real.

The slow crash means that in the West at least no-one actually starves to begin with. Increasing numbers of the middle classes drop down into the lower classes, however, and the lower orders naturally slip further down as they face even more severe pressures. Stronger economies become weaker because of this (with the more fragile ones simply collapsing altogether), while those nations which are still able to may engage in increasingly desperate wars to secure dwindling resources and to repel yet more migrants, thereby increasing the pressures on neighbouring countries less able than themselves.

At that point, with acute shortages not helped by disrupted supply chains and deflationary currencies, and growing numbers of unemployed and disenfranchised, it is all too easy to imagine the interruption or breakdown not just of law and order but of the key social services which currently we take for granted. Police and fire services, hospitals, banks and prisons, power and sewer utilities – all rely, as do the rest of us, on cheap fuel and therefore cheap power.

Take that away, and together with the erosion of the basic social order and social cohesion, what we think of as society suddenly ceases to function. Instead the frequency of food and other riots will escalate, and with the traditional authorities literally powerless to intervene against such numbers, pillaging and looting will eventually give way to killings by armed marauders.

When that happens much of what you hold dear becomes worthless or of no real use, just as many of the skills you currently take pride in or rely on become redundant. Your giant flatscreen television won't work, nor will your latest generation mobile or your new wifi internet connection. Similarly your once prized ability to handle office politics and your display of all the usual social graces will be as nothing alongside more fundamental attributes such as skill at hunting and maintaining a shelter – or indeed the defence of yourself and your family.

Then it is time to ask that most fundamental question of survival: fight or flight? Is home and its immediate environment a good place for you and your family to be? Is it somewhere which can shelter you and provide for you and protect you, somewhere which will be sustainable? Or do you need to be somewhere else? In short, should you stay or should you go, and what should you take along for the ride?

Chapter 12
A NEW NOAH

Anything saved will eventually be of some benefit.

Once the likely scale of the impending crisis has been acknowledged – and once one has accepted that vested interests and a continuing lack of any consensus will severely limit any attempt to reduce emissions to a level likely to make any real difference – it is time to stop and reflect on how we as individuals can best work to identify, safeguard and preserve elements of the world we know today. Clearly we cannot take personal, independent or even communal action to avert large-scale change. What we can do, however, is to devise and create arks – metaphorical as well as literal – in which to ride out the changes, while preserving key aspects of our lives and the ways in which we choose to live them today.

Biodiversity has to come first

Initially this means exploring ways in which we can safeguard a sustainable level of biodiversity. This has to be the first step – without it our own survivability is untenable – and so it is what we shall deal with first. Thereafter, what else we might choose to save remains open to discussion, and will be dealt with in the following chapters, although the fundamentals of what to save and how to save it are often very similar to those encountered when looking to preserve elements of the biosphere.

For example, the likelihood is that there are simply no right answers when it comes to deciding what to save – most obviously because we cannot know in advance to what depths humanity may sink, or indeed the full extent of the climatic and related changes that we shall have to deal with. Similarly it is impossible to ascertain with any degree of certainty where the greatest threat of destructive change will strike first or worst, or to what precise degree the changes will have an impact on the natural world and upon future generations.

For this reason anyone wishing to participate can really only act according to his or her best intentions – which actually, under the circumstances, is a rare freedom to cherish. He or she can also take comfort from the fact that the chances are that any particular decision in this regard will turn out to yield as good a result as a superficially more objective or 'scientific' strategy.

You personally, for example, might concentrate on saving an important component of the biosphere – supposing, quite reasonably, that in the coming decades this will be of some real benefit to mankind. A friend or neighbour may prefer to try to preserve some aspect of the knowledge or skills base which mankind has built up over thousands of years. Alternatively that same person may decide to work towards ensuring the survival of particular artefacts or works of art which they own or which are held in existing public collections. Finally, and somewhat more challengingly, the focus may fall on society itself, leading you, your family and friends to seek to devise viable ways in which to save valuable aspects of your own culture which might otherwise come under pressure or face total extinction.

Whatever is chosen, however, the likelihood is that anything saved will eventually be of some benefit, making it vital that you choose your scale of operation wisely. Saving something in its entirety may not be an option, no matter how desirable this might be. Instead would-be ark builders must keep a steady focus on achieving the possible, rather than attempting a bigger challenge that in the end can only fail. There is no point in acting otherwise.

Traditional conservation won't do it

In this last regard there is every reason to suspect that no matter how much effort we put in a substantial proportion of the biosphere will inevitably fall victim to global warming. Climate change is already directly threatening the survival of many different plant and animal species, and as time goes on the scale of this particular problem will become truly colossal as more and more species are forced to relocate towards the poles – and many find they are unable to do so within the time allowed.

Three years ago researchers at the University of Leeds put the figure for this at between 38 per cent and 52 per cent of the world's 5–10 million species, saying they would need to relocate to entirely new climate zones if they were to survive beyond 2050. Naturally the specific figures vary from one region to another, and from one species type to another as well. Thus when the Leeds research was published in *Nature* the proportion of known species threatened with extinction ranged from 10 per cent of Mexican birds through to nearly 70 per cent of South African mammals and nearly 80 per cent of its butterflies. Just over 20 per cent of plants in Europe are reckoned to be similarly vulnerable.

Fortunately we have the ability to help matters, but it will take a massive effort to achieve anything worthwhile and an almost total change in emphasis and policy. The relevant authorities need to be convinced of the need to act now, of course, since at present many nature and wildlife organisations – mindful of the impact of invaders such as Australia's rabbits and the kudzu vine in the south-eastern US – tend to be against the concept of taking plants and animals from already threatened populations and moving them elsewhere. Conservation, after all, literally means keeping things as they are, but while there are plenty of unknowns in this business we now know that sticking with the *status quo* is not going to be an option.

For example, and for a host of different reasons, traditional conservation efforts and sometimes quite unbelievable sums of money have been lavished on attempts at saving the more obvious trophy animals. Generally these tend to be the A-list celebrities of the animal world, such as the blue whale, giant panda and polar bear, and various eagles and ospreys.

Today successes with genetic coding offers us the possibility of preserving particular DNA sequences so that perhaps at a future date particular species could be successfully cloned. Even so, the reality is that many of these are almost certainly doomed: the giant panda, for example, not only subsists on a highly specialised diet and lives in a dwindling geographical range but also suffers from the triple disadvantages of tasting good, being clothed in a particularly striking coat which would just as readily make a very good rug, and being ridiculously easy to hunt down. At the same time it and many of its A-list cousins are already in such serious trouble without the additional burden of global warming that our efforts at conservation would almost certainly be much better concentrated on saving particular habitats and, where possible, whole ecosystems rather than these large and highly dependent if glamorous species.

The loss of top-tier carnivores, after all, will have a far less devastating effect than would the loss of many lower-tier organisms. Rising sea temperatures, for example, have already exceeded the thermal tolerances of some coral reefs, thereby destroying primary producers such as the photosynthetic *zooxanthello*e and affecting the entire ecosystem in these areas.

Think ecosystems, not individuals

This is because in order to save anything – plant or animal, wild or domesticated – it will be necessary to preserve its ecosystem if not in its entirety then to a very substantial degree. This will not be straightforward: it is extremely difficult even to identify an ecosystem in its entirety, let alone take the necessary steps to preserve it. (This has certainly been the experience of various experimental biospheres in the US, where it has proved far from easy to get a closed ecosystem to function fully without repeated or even continual human intervention.) Nevertheless attempts must be made to do this. There is, after all, little or nothing to be gained by saving a carnivore but not its prey, since any species which needs continual human intervention in order just to survive cannot be described as surviving in any meaningful or sustainable way.

For the same reason keeping a tree or rare shrub alive in your own back garden is not really an answer either, nor for that matter are the majority of zoos and safari or wildlife parks. Many of these may operate on a larger scale, and often from the best possible motives, but they provide only a partial solution and can perform little more than a temporary holding role in terms of the animals' long-term survivability. Conventional nature reserves are similarly ineffective: those that currently seek to secure particular ecosystems against outside influences are certainly likely to fail, and are already in effect like low-lying islands in an area where sea levels are rising. Most obviously they are doomed, however, because once the ecosystem in question comes under threat – from a change in its local temperature range, or because of too much or too little rainfall – the different species within it are bound to react differently, thereby upsetting the delicate balance on which their shared ecology ordinarily depends.

Some species of plant and animal will quickly prove more susceptible than others to parasites and disease that are encouraged by warmer weather. The creeping progression through northern Europe of bluetongue in sheep, deer and cattle is an example of this, as is the less publicised case of a native British species of coral which is already succumbing to the bacteria *Vibrio splendidus*. Other species will turn out to be more resilient, but even here it cannot be assumed that their survival will continue into the future without some external intervention – particularly if the climate continues to change over a long period and in the way that many climatologists expect.

To survive in any meaningful way the reserve would have to migrate to an entirely new climate zone, and do so in its entirety. Clearly this is something which it cannot do: trees, most obviously, have neither the mobility nor the necessary speed of generation to migrate at anything approaching a survivable rate. A species which takes ten years to mature, for example, and has a 50 metre seed range can effectively move at a rate of only 5 metres a year, which is clearly not going to be fast enough when climate change is progressing at its present, far more substantial rate.

Mobility is crucial

Because of this anyone wishing to preserve these species has to devise an ark which will effectively provide a means of transporting the ecosystem as a whole to another, more suitable place – which is to say one with a similar climate to the place which it has been forced to abandon but which also has a similar underlying geology, to make growth and survival possible in the long term.

While continual intervention is to be considered undesirable, the ark must be conceived in such a way as to provide a rolling solution, one which can be flexibly maintained while the process of climate change is going on, in the hope that, when things eventually settle down, sufficient species types will have survived for the relocated ecosystem to eventually settle down and once again become self-sustaining.

A good starting point here is to create seed banks, something which in this context may appear to be little more than zoos for plants although they actually differ quite markedly. This is because literally millions of different species can be preserved in this way, far more than would be possible in a zoo as we currently understand it. The most extensive of these – the Norwegian 'Doomsday Vault' which is hidden inside an ice-bound mountain on Spitsbergen – is reportedly large enough to hold all known varieties of the world's crops. It has also been expressly designed and constructed to withstand global catastrophes such as nuclear war or natural disaster, with seed collection being organised by the Global Crop Diversity Trust. The final number of species thus 'saved' will be determined by the countries wishing to use it, and being prepared to pay the costs for doing so, but should still be considerable.

Such ark builders also need to avoid the mistake of being highly – which is to say dangerously – selective when it comes to deciding what to store and maintain. What might be termed a weed or an unpleasant virus or dangerous bacterial strain, for example, might be allowed to slip into extinction simply because no-one realised until it was too late that it was actually the lynchpin of an entire ecosystem.

A perfect illustration of this came as recently as August 2007 when two researchers at the University of Zurich demonstrated that many vital ecosystem services – including water purification and CO_2 storage or sequestration – depend on a far larger number of individual species than had been thought hitherto. That is obviously quite a concern given the growing threat of water shortages globally – not to mention the increased interest in CO_2 storage possibilities – but particularly so when one considers that at the time of writing an incredible 34,000 different plant species were already reckoned to be under serious threat.

Of course it is also necessary to establish how these seeds will cope under conservation conditions, since our knowledge of this is still very slight. Research into the behaviour of seeds has so far examined just 4 per cent of flowering plant species, and already it has been ascertained that many of them are simply unable to tolerate being dried out for storage in the conventional manner. Most oaks, for example, quickly die if they are dried below about 40 per cent moisture content.

Assisting migration

New Noahs will also soon realise that it is not just those species that are literally rooted to the spot which come under pressure as a consequence of climate change. Many species that can physically move with the weather will also run into trouble without some careful (if temporary) human intervention, as changing temperatures and patterns of precipitation affect not just their own habitat but also that of the food – be it plant or animal – on which they depend.

Here the risk is that as the effects of climate change begin to take effect the habitat in question, be it natural or conserved and like the aforementioned nature reserves, will become in effect a low-lying island surrounded by rising sea-levels. Some species will doubtless be able to move to higher territory (as it were) but even these, once they reach the summit, may find themselves stranded if this too proves eventually to be insufficiently elevated above the waves. There may, it is true, be more appropriate environments or 'higher islands' relatively nearby – but many of the threatened species cannot physically leap from one to another.

Immediately one can see the importance in Darwinian terms of acquiring mobility, and the huge advantages that are enjoyed by birds such as the Arctic tern, which circumnavigates the entire planet and flies tens of thousands of miles a year, or even the little swift which lives almost entirely on the wing. These – in what a Darwinian might call 'the survival of the fleetest' – display the last word in mobility, arguably more so even than fish. By contrast large land mammals and the like will do rather less well without help. This can be seen in their numerous historical extinctions – particularly in hot, dry, isolated Australia, and indeed in the more recent example of that country's kangaroo population, which is estimated to have been depleted by around 50 per cent during the Big Dry.

Helping such species will involve the creation of so-called migration corridors – another ark – and this must become an immediate priority. If these are carefully configured some species will be able to safely move through them unaided. Others will need to be assisted, however. In the past evolution gave species the time they needed to make the change, but, as we are now seeing, the speed of climate change will deny many of them the hundreds or even thousands of years they would ordinarily need to move these relatively long distances.

Of course even trees can and do migrate already: evidence from fossil pollen indicates that some North

American tree species might have moved at up to 1 kilometre a year at the end of the last ice age. Such a speed may not prove sufficient, however, leading *New Scientist* to suggest that one answer could be to 'FedEx struggling species to comfortable new locations'. In fact biologist Dov Sax from Brown University, Rhode Island estimates (he says conservatively) that around 10 per cent of all species will need help in this way if we are to avoid losing them completely.

To achieve this conservationists will have to overcome their usual scruples and to accept that, just as (taking a northern hemisphere example) they personally might support the notion of rescuing their own local flora and fauna by sending examples into new territories further north, they must also accept alien species coming into their own environment from the south. The good news here, at least as far as they and their associated conservation bodies are concerned, is that in general the species likely to be helped in this way are not those – such as rabbits and fast-growing weeds – that in the past, thanks to their combination of rapid reproduction and dispersal, have tended to overrun their new habitats, thereby causing massive additional problems.

Consider species substitution

Another solution to the problem of conventional nature reserves – but one which also runs counter to established conservation practice – is species substitution. Because many natural ecosystems will fail if they remain stationary and unchanging, consideration might be given to introducing species that will be able to cope with, for example, hotter summers and less regular precipitation.

The least controversial option here would be, where appropriate, to breed new strains which have been selected for their survivability under the expected new climatic conditions. These, of course, will themselves be far from static, but the Nature Conservancy in Virginia is already developing this sort of strategy by encouraging more heat-resistant varieties of native plants to grow, and also by identifying sites – such as those by large bodies of water or on northern slopes – where the impact of climate change might be lessened.

Even if this is successful, however, conservationists must also consider importing wholly alien species where appropriate examples can be found to replace native species facing probable or certain extinction. Should we in the UK, for example, already be looking at, say, the Dordogne in France to see if there is anything growing there that we could reasonably plant here? Species of tree which mature faster and cast seeds further, for example, and can survive and thrive with far less water.

Again, hardly surprisingly, this last recommendation goes completely against all conservation efforts so far; but this is only because conservationists' objectives traditionally have been quite different to the ones we now have to pursue. Doing it this way is nevertheless bound to be a somewhat challenging notion, though, and not just because there have in the past been several disastrous experiences of introducing alien species – sometimes deliberately, at other times as a result of accidental release – only to see them overrun fragile environments or wipe out these native rivals. The reality remains that the rules are changing fast, and conservationists everywhere need radically to rethink their priorities and preconceptions and to alter their methods accordingly.

Wholesale relocation and recreation

For the same reason consideration must also be given to the wholesale relocation of some other more exotic ecosystems. Coral reefs, for example, are well known to be highly sensitive to changes in temperature and are already coming under very severe stress in the southern hemisphere. Here again, being island-based, they will need considerable human assistance if they are to survive. Perhaps entirely new reefs could be seeded in those waters which are becoming warmer: the Canaries, perhaps, rather than the Scillies, given the projected possible failure of the Gulf Stream.

There is also the problem of what to do about bogs, salt marshes, mangroves and other coastal margins. All of these face something of a double whammy, of course, being badly affected by changes in both sea level and sea temperature while their potential for natural (unassisted) migration is likely to be negligible. Here perhaps, and certainly when it comes to the pressing example of commercial crops, it will probably be necessary to undertake a programme of terraforming, bringing artificial assistance to bear on the creation of an appropriate landscape.

Both wild and domesticated plant species, after all, require the right soil as well as the right climate, and while in theory it might seem logical for, say, the US mid-west corn belt to gradually migrate northwards to Canada as a way of coping with climate change, the underlying geology may well not permit this to occur. Good fertile topsoil is a crucial requirement for the production of around 97 per cent of the world's terrestrial food supply and, as mentioned previously, glacial sands, gravel and scored rock surfaces are no substitute for the type of rich mature, fertile soils upon which efficient agriculture depends.

The soil found in the English fens and other agriculturally productive regions is the result of many hundreds of years of natural activity, with weathered rock and decomposing plant and animal materials being gradually broken down and mixed by a combination of plant roots and microbes, invertebrate activity and fungi – and of course the weather. However, and as we have already seen, much of this is now under threat. David Montgomery, a University of Washington geomorphologist, estimates that the world's soil is being eroded at least twenty times faster than it is being regenerated, to which must be added the additional pressures of changing climate, inundation by seawater and simple exhaustion, as a consequence of intensive agricultural practices.

As a result, and without time and Mother Nature on their side, the likelihood is that farmers in the near future will have to spend many years planting and encouraging non-productive, fertility-boosting species to grow before any more useful, truly productive crops can be raised in their place. Of course given sufficient infrastructure science may also offer some assistance here, for example by mixing organic waste such as chicken litter and garden rubbish with fly ash (a byproduct of coal-fired power stations), or indeed ordinary clay mixed with the leftovers from industrial cornstarch production. A variety of different combinations have already been trialled by US, British and Australian academics, and although some have proved too toxic to be considered for food production the underlying principle seems proven, and it may in time be possible for our children and their descendants to manufacture safe, fertile topsoil using relatively simple technologies.

Chapter 13
FIND YOUR TRIBE

Centre your efforts on the idea of an ark as a genuinely defensible space.

When it comes to safeguarding the social dimension, key decisions need to be made and put into effect early on. More significantly, as Gus Speth, former President of the World Resources Institute, observed some years ago, the impetus has to come from private individuals because the politicians have already let us down and will do so again. If citizens don't take the helm, argues Speth, we'll lose the fight – it's that simple.

Therefore at some point we need to ask: am I seeking to save just my family and friends, or a larger definition of community, perhaps even an entire nation? It might also be useful to consider the behaviour of native Americans when they think in terms of society – an indigenous people who are traditionally said to think seven generations ahead.

That said, on even the smallest and most personal scale – that is, when it comes to safeguarding your family or a small group of likeminded friends – the situation looks and sounds intimidating, because whatever else you decide to do you must centre your efforts on the idea of an ark as a genuinely *defensible* space. This means of course that you need to consider whether or not you personally know how to defend yourself against any sort of physical incursion – aside from ducking and running, which might quite naturally be your first inclination. In other words, you need to ascertain whether or not you have any offensive or defensive skills which can be brought to bear, and some idea of where your new home might best be situated.

Identify your real *community*
Inevitably the answer to the latter will be a function of how large or small your chosen group happens to be. If it is less than ten, for example, perhaps a conventional nuclear family with the odd relative in tow, you could in the first instance do worse than get a boat. Offering numerous key benefits – among them mobility, some level of sustainability and a degree of isolation which can itself offer a first level of defence – such a literal ark is also usefully modest in terms of its desirability to others. In fact it offers aggressors very little, which is good, little more that is than the chance of a pyrrhic victory should they wreck your vessel while trying to take it over. By contrast, a superficially more attractive base, such as a well-stocked filling station perhaps or a supermarket full of canned goods, would merely render you and your family Target No. 1 among

a wide variety of marauders as keen as you to secure a ready source of food and fuel. The key here, of course, is sufficiency rather than over-supply: aim any higher than that and you simply become a magnet for predators.

Partly, but not solely because of this, much larger groups present rather larger problems – although these too are by no means impossible to resolve. That said, and as demonstrated by examples such as Pennsylvania's historic Amish community (and on a smaller and more recently established scale any number of North American survivalist clans), the initial focus must be on identifying and establishing a group of individuals that is like-minded and emphatically monocultural. That is, one in which the individuals share the same overall objectives but between them are still able to demonstrate a proper diversity of useful skills.

You need to find your tribe, in other words, and be aware that this may no longer be the group with whom you have traditionally allied yourself, such as your country, your immediate community, a racial or religious grouping, or followers of a particular political party, religion or even sports team. Unfortunately the reality is that too often relationships within this sort of group are not actually cooperative and open, but often competitive, highly coercive, exploitative and even abusive – none of which will serve you well in the future. In their place successful survivors are more likely to be members of more discrete, more obviously multi-skilled, well-balanced and cooperative groups – the make-up of which may not at the moment be especially obvious.

Secure a redoubt

Thereafter, once a tribe has been identified, the first major task is to find and secure a defensible space of its own. An island, perhaps, or a mountain, but of course only one that is able to provide a subsistence level of the necessary resources. At the very least this means good soil and clean surface water for

irrigation and consumption; that is, a spring or stream you can
drink from, or perhaps a well which can be accessed without
power. Dirty surface water would be fine too, incidentally,
providing the water itself can be filtered and cleaned using,
say, sand and reed beds. Of course rain water will also serve
the purpose, providing it falls in sufficient quantity and your
community has the means to capture and store it in the event
of seasonal droughts.

Any such groups may also like to consider allying
themselves to a larger monoculture. That of Japan, perhaps, or
even China, New Zealand or Argentina – although here it is
essential that your members can blend in with the larger
group: as difficulties in society increase, scapegoating is likely
to be on the rise in any society which comes under pressure,
and nobody wants to look and sound like an outsider. Newly
exploitable regions may also start to become attractive in this
regard, for example parts of Scandinavia, Canada, Alaska and
even the lower latitudes of Siberia, if the right combination of
climate and soil can be found to afford migrants to these
regions a sustainable lifestyle.

Here again the key word is sustainable. Once any
group's redoubt is established it must be designed along
purely subsistence lines rather than overly productive ones.
A sufficiency of food and water should therefore be not just
the starting point but also the end point since, as we
observed previously, storing wealth or capital is merely to
send an invitation to invaders, desperadoes or bandits –
particularly in an environment where there is no longer any
effective law and order. For this reason alone many
marginal, subsistence communities in sub-Saharan Africa
have long recognised that under such circumstances –
which in our own case means living in what could quickly
become a failed state – it is far better to measure wealth in
terms of skills rather than things, as the latter can and will
be stolen.

This is not to say that skills cannot be stolen – a skilled individual can obviously be enslaved by someone with more power who lacks their particular talent. But if that happens then he or she will at least remain alive, and for this reason 'better fed than dead' might be a useful motto for the future. That, and a commitment to avoiding the tall poppy syndrome – aiming to be a smart, silent peasant – may turn out to be the most useful strategy for survival once those basic skills are safeguarded

And it is of course vital that they are safeguarded. The preservation of a recipe book, for example, will serve no purpose unless the future reader also knows how to cook; similarly it is not enough to save a match, but instead one must save the skills and the knowledge required to make more matches when the first is gone.

Think skills, not possessions

The initial focus within your tribe or grouping must therefore be on identifying and securing the right skills, which is to say those skills which are self-contained and continually prove their worth day to day. There's a saying, isn't there: have a fish, eat for a day; learn to fish, eat for a lifetime; and clearly in the future (as in our distant past) the most valuable skills are going to be the most useful and most practical ones. These are the ones concerned not just with building (or improvising) shelters from salvaged objects and found materials but also finding or catching and preparing food. Finding water and making fire will be key as well, together with some more modern, more technical ones such as medical diagnosis, even basic surgery, together with navigation, organic gardening, composting and animal husbandry, the more practical aspects of physics and chemistry, and of course metalworking and mechanical repair.

As one blogger puts it, 'as the fifteenth century had Renaissance Man, we're going to have the Post-apocalypse Man or Woman, someone who can fix a bicycle, tan a hide, set a broken bone, mediate an argument, and teach history'. Another way of looking at the issue might be to consider that, in the future, the craftsman-glassmaker is likely to be a lot more useful to us and our community than the erstwhile managing director of a glassmaking company.

Thereafter, and within that particular limitation, it will be necessary to rebuild our lives as hardworking members of these new smaller, more sustainable communities. Sustainable because their component individuals will have discovered in good time that they need each other in order to survive. Sustainable because, forming what is essentially a fairly simple, almost tribal social system, they will be able to work together to grow, gather or hunt and prepare their food, to look out for their fellow members and – when necessary – to arm themselves against outsiders.

Fortunately, if somewhat bleakly, starvation will have taken care of most of the population before too long, particularly those mavericks and so-called lone wolves who – in a simpler world that requires greater levels of cooperation than the present one, and demands the true sharing of resources – will mistakenly believe that they can best survive by travelling light and alone and by taking what they need from those who already have it. In reality the survivors will be those who recognise the true meaning of sustainability: finding and securing a food supply which meets a whole set of new criteria, namely:

- ♦ Accessibility/Proximity – a good food source will be one which is as close to the community as is practical.
- ♦ Healthy – its availability should provide for a balanced diet without the need to call on the resources of other communities, which in the short term may well seem themselves as rivals.
- ♦ Fairness/Non-exploitation – on even this small scale there must be fairness and cooperation between producers and consumers.

♦ Environment – with such limited resources it is essential that the food not be obtained at the expense of the local environment.

♦ Bio-diversity – for this reason maintaining a diversity of varieties and breeds within the immediate locality will be crucial. (Animal welfare standards must similarly be maintained.)

Take something with you

Finally it is incumbent on everyone to remember to take something of the now with them for our tomorrow – that is, to build an ark of their own – because without this, simply surviving will in the end turn out to be just another failure.

Chapter 14
KNOWLEDGE, TRADE AND CULTURE

Our increasing reliance on the technically sophisticated quickly becomes a hindrance rather than a strength or an advantage.

Perhaps surprisingly, when it comes to considering the preservation of the man-made rather than the natural, many of the challenges we face are in essence much the same as those explored in the previous chapter on biodiversity. Thus if one substitutes 'infrastructure' for 'ecosystem' it quickly becomes apparent that many of the problems and obstacles we need to overcome in order to preserve many of the more detailed aspects of our civilisation – that is, what one might consider to be our civilisation's greatest historical and technological achievements – are very much the same as those involved in safeguarding a meaningful proportion of the earth's biosphere.

Most obviously we find ourselves in a situation where, once again, few if any elements of our daily lives can be treated in isolation and must instead be considered as part of a whole. And unfortunately we find upon doing so that our increasing reliance on the technically sophisticated and scientifically highly advanced very quickly becomes a hindrance rather than a strength or advantage.

Hi-tech more threat than cure

Highlighting a very obvious example of this – and, more significantly given the context of this book, underlining society's failure so far to engage with the challenges posed by climate change – in July 2007 the UK National Archives announced what it referred to as a landmark deal to rescue future historians from what it called the digital Dark Ages.

From that announcement it emerged that the vast majority of documents in the archive, much of it dating back many hundreds of years, are handwritten and therefore still legible at least to those skilled at unravelling ancient texts. But increasingly, and here's the problem, many important documents are being archived in the form of computer files, sometimes because this is how they were created in the first place but also because it is more space-efficient, and also makes it easier and safer for the valuable documents to be retrieved and studied.

Unfortunately, though, such files run the risk of becoming unreadable as successive advances in computer technology render existing computer retrieval and storage methods obsolete. Therefore, to get around this not insignificant problem, the National Archives has teamed up with the British Library in London and Microsoft in the US, their combined aim being to create a new, more powerful and fully flexible computer system, one which will be able to unlock what are known as 'legacy files' in order that they can be seen in their original format, regardless of any changes made to computer hardware and software over the coming years.

The need to do this is certainly there, since data or knowledge stored in the conventional manner on a floppy or hard disk ordinarily has a lifespan of only ten years or so. In itself, of course, this presents no problem whatever as long as the technology and expertise exist to ensure that both the files and their means of storage are regularly renewed and refreshed. But if we lose the ability to do this – and it is highly likely that in the post-carbon era we shall – there is a very real risk that the totality of man's knowledge could begin to decay very rapidly.

When this happens the reality is that even these 'rescued' legacy files will be effectively useless. There will, after all, be no point in preserving a computer disk or hard drive unless one also has the means to interrogate it. And indeed no sense at all in retaining the appropriate model of computer unless one has managed to secure the necessary infrastructure – which in this case includes not just the means of powering that computer but also a means of repairing it when it goes wrong and replacing components as they wear out or lose function.

Internet: a single point of weakness

This particular problem is amplified by the existence of the internet, the greatest benefit of which has clearly been to make more information accessible than ever before, and to many more people than at any time in history. However, while it provides what is rapidly becoming almost universal access to the totality of human invention, intelligence and ingenuity, the internet is at the same time a uniquely dependent and uniquely vulnerable entity. Most obviously it is dangerously reliant on a wide range of different technologies, which means that we are too. The way in which all this information is stored and distributed means that every user, more than a billion of us at the present time, is now utterly dependent on this uniquely complex infrastructure – and becoming more so as every day passes.

This is something which needs to change if we are to retain the vast store of knowledge which we can call on today at the click of a button. At present we can do this so easily that there is, for example, no longer any need for us to keep local copies of any of the information stored online – be this a work of literature, a legal document, or even a set of instructions about how your new camera or mobile telephone works (or for that matter your computer and its internet connection).

Consequently none of us does. Instead, we can call up the relevant pages whenever and wherever we please, an arrangement which for now at least is not only quick and efficient but also highly convenient and, of course, apparently quite 'green', in that great swathes of forest no longer need be cut down in order to produce (for example) millions of instruction booklets, which increasingly no-one reads.

But while the internet is many senses a genuine technological triumph, and while one sincerely applauds such initiatives as www.gutenberg.org – an organisation which has so far digitised tens of thousands of rare books and documents and made them much more widely available – this method of data storage is nevertheless also dangerously *un*diverse. On the one hand it increases the availability of information, but at the same time, if somewhat paradoxically, it also centralises an immense body of crucial information, leaving us all exposed to the risk of a single technical failure – or, in the future, a complete and possibly irretrievable technical breakdown. Thus major search engines such as Google or Yahoo, for example, could very quickly turn out to be this massive entity's Achilles' Heel, a single point of failure within what is otherwise a truly global network. Such companies, after all, don't simply provide the means of access to the information which makes up the world wide web but also function as the index to it all – an efficient and highly organised filing system without which we would be unable even to establish for certain where a particular piece of information is, never mind gain access to it.

And of course increasing computerisation of this sort is just one example of the potential dangers that are inherent in the high levels of specialisation which we now take for granted. Just think about a simple disk drive for a moment. It looks simple enough, and these days many of them cost very little. But actually such a device requires not just a lot of very specialised components to make it work but also extremely high levels of precision when it comes to its design, assembly and operation, and a surprising number of very specific materials when it comes to maximising its overall efficiency.

Substitution and simplicity hold the key

In real terms what this means already is not just that most of us can't fix the disk drive when it breaks, but also that even the experts who do know how to fix such equipment still require the support of an entire industry to make it and to keep it running sweetly. Not just the factory which manufactured the disk drive in the first place (and indeed the disk inside) but also a whole array of first, second and third tier suppliers, a mining industry able to extract the appropriate materials, processors to make them industry-ready . . . and so on and so on right down the line. Computers, in short, are part of an unbelievably dependent infrastructure – and that is a very real danger to us all.

There are some substitutions which could be made to alleviate some of the likely problems – but all of these have cost implications in terms of their efficiency and functioning, and most will be only very short term solutions. Given sufficient power to refine it one could use aluminium instead of magnesium, for example, or replace the tantalum capacitors in our mobile telephones – a uniquely ubiquitous form of computer, hence the tantulum running out – with some other semi-suitable metal. True, the capacitors will need to be four times the size as a result, but as tantalum reserves dwindle to nothing something will have to give, and in the end the computer as we know it will most likely be it.

The short-term nature of such solutions, however, means that there is already an urgent need for people to choose and isolate the knowledge they wish to preserve and – starting now – to begin to create local caches of this information, using a medium which will be both enduring and less dependent on a complex, fragile infrastructure.

Fortunately many academic journals already provide just such a function for a great body of highly specialised information. This is fortunate, and not simply because the knowledge they record is on the whole significant, peer reviewed and of a reliably high quality. It is fortunate also because, while the contents are invariably digitised and available to subscribers and others via the internet, hard copy of it also exists and tends for obvious reasons to be stored in a variety of secure, appropriate and cherished environments such as university libraries and national scientific archives.

That said, of course, paper may not in the end prove to be the most durable medium either. It too decays with age, suggesting that in the future perhaps we will need to see a return to a sort of 'scriptorium' system, wherein teams of individuals work as medieval monks once did, copying out pages and pages of information in order to safeguard it for future readers.

At least one can again take comfort from the fact that, while choosing precisely what knowledge to safeguard may not be easy, each of us has the freedom simply to *save whatever we like*, because the chances are this will be as good a means of deciding what to save as any other. Also, with a population of more than 6 billion and rising, the likelihood is that what in the end turns out to be the right sort of knowledge – the most useful or the most valuable or the most enduring – will be saved somehow by someone somewhere. Not, mind you, that this should in any way contradict the view that the so-called wisdom of crowds is anything but an extraordinary and enduring popular delusion. . . .

Survival more likely for simpler societies

Saving our culture, however, will be much harder if by culture we mean the complex social structures we value within our existing communities and nation states. These are structures that in many cases have been built up over many thousands of years, and deciding what to defend and how to defend it presents a much more challenging scenario. This can be seen when we look at how even the most stable, well-regulated and most liberal societies in the West can be and have been rapidly and substantially undermined by relatively minor incidents.

The sudden destruction of New York's Twin Towers, for example, was obviously a unique, shattering and horrific event – and not just for the citizens of Manhattan or for those personally involved. Yet in purely numerical terms the scale of 9/11 is dwarfed by the numbers of those killed each year in, say, road traffic accidents in the UK let alone across the whole of the US, or indeed of those whose lives and communities are daily threatened by climatic upheavals such as the expansion of the Sahara. Even so, the economic damage arising from this one single criminal outrage was and continues to be multiplied several-fold by the actions of politicians and others who are attempting to ensure that something similar will never happen again.

The reaction, in other words, is in many ways – financial, social, economic – arguably more damaging that the original incident. Most obviously the whole culture of the West has shifted and continues to shift as part of the so-called war on terror. Typically, changes to the way we live our lives have included increased surveillance, much greater restraints on the movement of individuals, and a welter of actual and proposed legislative changes which threaten to severely compromise or in some cases even eradicate completely many of the freedoms which have been won over hundreds of years.

Given such a scenario it is easy to see how the kind of seismic shifts likely to result from climate change are bound to upset the dynamic of even those societies that seem at present to be the most secure. And once this happens monocultures such as small island communities would appear to have the best chance of survival – assuming, that is, they are sufficiently elevated to avoid inundation by seawater – and certainly a better chance of prevailing than many much larger and more prosperous countries such as those of Europe and the US.

They have less to preserve, for one thing, making the initial objective much easier to achieve. For another, their relatively small size means their citizens are more able to reach a consensus and to act together. An example of this might be those small island communities which have already chosen to preserve their own culture by more or less rejecting large-scale tourism. This is often done on the grounds that – while development of this sort will almost certainly bring opportunities and wealth to a few on the island – for most individuals it is simply a means of volunteering for wage-slavery and localised environmental degradation.

People living in marginal lands that remain marginal are similarly likely to be better able to protect their culture, if only because there will be little economic benefit to anyone invading or taking over their territory. Here small truly is beautiful, and if you have little (as we have seen) you have little to steal. Of course this does not alter the reality that many of the better known marginal lands – the high Arctic and the rainforest being two of the more obvious ones – are certain to suffer badly as a consequence of climatic change, and regardless of what their indigenous peoples decide to do.

Unfortunately the other reality is that our advanced western cultures face a challenge that is every bit as stiff: to mount an effective defence without destroying the very aspects of these cultures – such as democracy, and a multitude of smaller, individual freedoms – which make them so desirable in the first place.

In conceiving arks of all kinds one needs to consider the possibility, for example, that a powerful nation might seek to preserve its own national culture by using its overwhelming military strength to defend its access to whatever resources it considers necessary, if not to eradicate potential rivals. Of course these days such a brutal, simplistic notion – truly red in tooth and claw, of nationalism versus liberalism – sounds far fetched. It was after all very much the rationale behind Japanese aggression in the 1930s, behind that of the Nazis' own plans for expansion in the same decade, and indeed for many other empire builders throughout history. Nevertheless it is true that anyone who sees in recent Middle Eastern adventures an American attempt to secure its oil supply – and very many commentators do – has only to make a relatively small leap in imagination to picture a scenario in which the US seeks actually to control that oil and thereafter, as supplies diminish, to control an ever larger proportion of that same crucial resource throughout the region.

The threat to trade
Having touched briefly on the virtues of simpler, less dependent technology and networks in the previous chapter, it will come as little surprise to discover that the complexities of modern international trade – an important defining characteristic of any fully developed society as well as a crucial

engine of continuing development – mean that this too will be seriously disrupted during the coming years.

This will happen not simply because some industries will eventually become unsustainable – although of course the nature of many means that they will. Nor will it happen just because the bulk of such trade at present relies entirely on fossil oil (to fuel the shipping, aviation, road and rail transport infrastructures which are needed to move the goods around), although the risk of this is high as no viable alternative has yet been explored or established.

It is inevitable too that climate change itself could in time bring to an end the whole process of globalisation, its collapse taking place gradually as individual nations look increasingly inward and seek to retain their own scarce resources or – even more worryingly – to obtain control over those resources located elsewhere. Naturally the likelihood of the latter will increase dramatically in the face of depletion – we are already seeing this in various quasi land-grabs in the Arctic and Antarctic – particularly as demand for those resources rises as a result of growing populations, and indeed the influx of environmental migrants fleeing rising seas, increased desertification, agricultural collapses and so on.

The obvious key to all this is resource scarcity. Not just of the fossil fuels required to move goods but also of the tradeable commodities themselves. Existing models and systems of global cooperation and trade, after all, are built on the basis of a resource-rich world with willing sellers and willing buyers. The latter will always exist, but willing sellers will gradually retreat as various key resources become more and more scarce.

Indeed, Leon Fuerth of George Washington University, one-time national security adviser to former Vice-President Al Gore, sees scarcity dictating the terms not just of trade but of all future international relations. Foreseeing trouble as early as 2040, Fuerth claims that richer countries will begin to 'go through a thirty-year process of kicking people away from the lifeboat' as the world's poorest communities face the worst consequences of environmental change.

Unsurprisingly this is a process he describes as 'extremely debilitating in moral terms', and he is not alone. Based on even the IPCC's middle-ground estimate – a 1.3°C increase in average global temperatures and a sea level rise of just 9 inches – John Podesta, Bill Clinton's former chief of staff and now president of the Center for American Progress think tank, predicts a similar scenario, one 'in which people and nations are threatened by massive food and water shortages, devastating natural disasters and deadly disease outbreaks'.

Against such a background it can readily be seen that the whole notion of happy seller/happy buyer will collapse, something which can happen very quickly – as we have seen (albeit on a minor scale) in the way in which the relationship between oil consumer and producer nations was fractured once the politics became explicit in the early 1970s, when OPEC gained the whip-hand. Trust is important in trade, but what, one wonders, will happen to this when oil gets really scarce, or when we 'go green' – and when the present happy buyers and sellers are replaced by desperate ones on one side or the other?

Better to barter

Against such a depressing background one small note of optimism is perhaps sounded by a multitude of archaeological evidence that demonstrates even early man was able to partake in some form of trade. Admittedly, reinstating something akin to this runs against the recommendation in Chapter 13 for new communities to choose subsistence over surplus (in order to avoid hostility from rivals or roaming mavericks). It is certainly important that our new, smaller social groupings concentrate on building their future on the basis of self-reliance.

Trade on even a modest scale will prove vital to our future, however, although for obvious reasons any attempt to rebuild something akin to what we have now will quickly prove futile. Instead, and in the absence of any real trust between communities, trade between them will need to start on the basis of barter rather than sale and purchase. The advantages of this are obvious, as is the initial need to operate these trades on a bipartisan basis rather than using a third party to move the goods from one community to another, which would quickly expose the system not just to theft by the transit operators themselves but also to pirates.

There is no need to predict this latter aspect, of course, since we can already observe it happening now in several so-called 'failed states', and in many other regions and territories where what we would recognise as the rule of law has either never existed or more recently simply broken down. Here too we can see the robustness of this sort of simple trading system, and the fact that largely self-reliant communities (particularly those with a strong warlord or local ruler) can resist incredibly strong external forces – think Afghanistan – and survive what appear to be overwhelming onslaughts from the outside world. Clearly if we wish to preserve our own community we must learn from such examples.

Art and artefacts

This is a necessarily short section, and a relatively easy one to write because all it needs to say is this: save what you like.

Really that's all – not because I don't care about such things or consider that they are not worth saving, but rather because with a little thought and planning the portable and the physical are surely among the very easiest things to save. It is also the case that a vast number of artworks and artefacts are already well on the road to preservation, being the property or in the care of those giant proto-arks – the great museums and art collections of the world – although it is important that we don't bury the collections now or hide them away, as this effectively constitutes theft from the current generation.

There is, after all, a cost to all this, so any would-be saviours need to consider the impact on others and ask themselves whether the cost is worth the price. Ark-builders might also like to consider the value of the traditional time capsule, a biscuit tin or similarly robust, sealable container of the sort which has been buried everywhere from the foot of Cleopatra's Needle to the Blue Peter Garden. Combining simplicity with utility, the construction and concealment of this sort of ark falls well within the capabilities of the individual.

That said, with objects and artworks it is also important that we don't simply assume that someone else will take care of business and that we take steps to avoid unnecessary duplication of effort. With a painting the latter is easily done, for two people working apart obviously cannot save the same picture. But in other cases there may be multiple copies of a particular artefact or object: it's obviously best to save something else instead.

Fortunately collectors already frequently operate at different levels within the same arena, so that one passionate record collector might have spent years amassing every one of literally thousands of different Bob Dylan recordings – live and studio, official and bootleg – while another might simply have set out to collect every No. 1 single ever, regardless of the artist, song or indeed artistic merit.

Here again it is important to consider the infrastructure. Put simply there is no value in anyone preserving a cherished record collection if nobody has taken care to preserve the equipment on which to play it. Consider also that obsolete technology may eventually come into its own: a wind-up gramophone may lack the high fidelity of a CD or MP3 player, but it is a far less dependent form of technology and one that is relatively easy to repair and maintain. (This is why, of course, relatively unsophisticated machines such as the original Land Rover and the Russian AK47 assault rifle continue to find friends around the world, while in the West the armed forces have come to rely on increasingly complex and astonishingly expensive equipment – which is often difficult if not impossible to maintain in the field.)

Defending reason against religion

Perhaps fortunately – although by no means necessarily – another key aspect of our culture, namely religion, is far easier to preserve and as a consequence is likely to fare better than it does at present. Religious organisations, such as the Catholic Church on a global scale and more locally the Church of England, are already natural arks. Adversity is known to be good for religions, and as we have seen time and again oppression and other pressures make fine recruiting sergeants to the cause. Evidence of this is not hard to find: just look at the resurgence of Islam, now that Muslim populations in many different countries perceive themselves as being under attack.

The tragic effects of this are there to be seen, and unfortunately the additional pressures brought about by climate change will serve only to accelerate this effect. What's more, they look likely to do this even without the unwitting assistance of the likes of the Right Reverend Graham Dow, Bishop of Carlisle, who went so far as to blame the 2007 summer floods in England on mankind's behaviour, suggesting that the death and destruction visited on the north of England in particular was a judgement from God, following the Church of England's acceptance of homosexuality.

It is conceivable, for example, now that mankind's responsibility for climate change is generally acknowledged, that in certain sectors of the population aspects of modern life such as technology, progress, even science itself could become something of a scapegoat for these latest difficulties. The obvious effect could be to cause that old dichotomy to resurface – religion versus reason, the idea that it was the Tree of Knowledge which removed us from Eden – and for there to be something of a backlash against scientific knowledge and indeed governments if they are shown to be ineffective in dealing with the resulting crises.

Were this to happen it would leave the way open for religious functionaries to once more become very powerful individuals, and for churches themselves to experience a great resurgence of their power. Religion, indeed, far from being in danger, could become instead a danger in itself – and at least from a rational, scientific point of view that has to be a worry.

Chapter 15
THE WASH TIDAL BARRIER

My own ark has many potential benefits.

So, faced with all this, it is fair to ask what is the author doing? I'm writing this book for a start, as I believe it to be the best way to kickstart a proper action plan to deal with the coming challenges, rather than just sitting around debating whether man is to blame for the changes, or whether the changes we are seeing are just part of a normal cycle of nature. Normal, that is, but still a problem of terrifying proportions with the global population at its current level.

On a smaller scale, I am using the train more often,
I have given up my day job and bought a farm which is on
higher ground than the surrounding countryside, and I am
discussing the future and what it holds with my children
because I recognise that the challenges facing them are
going to be more substantial than those facing my own
generation.

I am also promoting a much larger project, namely the
Wash Tidal Barrier, having established that what I value and
wish to protect is the view from, and environment
surrounding, my home.

As we saw in Chapter 10, and at a time when we are all
of us becoming acclimatised to what until very recently was
still regarded as strange or freak weather conditions, it would
take just a single big storm in the North Sea to coincide with a
spring or otherwise exceptionally high tide for some 300,000
hectares of Grade 1 English agricultural land to become
completely inundated with seawater.

Were this to happen the damage and long-term
implications would be immense. Not just because the area in
question, namely the low-lying fens of Cambridgeshire,
Lincolnshire and Norfolk, comprises some of the richest and
most productive agricultural land in the whole of Europe;
there is also the local population to consider – currently put at
around half a million – and the literally billions of pounds
worth of infrastructure in terms of people's homes, businesses
and other public and private property.

In late 2007 just such a threat surfaced, with the waters
eventually receding after reaching just 8 centimetres below the
current defences. Had they breached these defences it is of
course true that the floodwaters would have subsided relatively
quickly. Desalination would have taken years to complete,
however, and the cost would have been huge in terms of both
the clear-up operation itself and the resulting loss of crops and
production.

It can be seen then that, even without freak events such as this one, climate change and resulting sea-level rises between them represent a real and imminent threat not just to the coast of the Wash but also to the fenland environment – an ancient but nevertheless essentially man-made landscape. Initially these breaches will be occasional, perhaps every few years or so. But over time their frequency will increase, requiring any land contaminated by the water to be chemically restored to full productivity, an operation which will itself take years to complete and be extremely expensive.

In fact, with sea temperatures already higher than in recent times, and currently rising by 0.4°C per decade, and sea levels forecast to rise by at least 1 metre by 2100, it is already abundantly clear that existing coastal defences will soon be wholly unequal to their task. It is apparent too that raising the height of these defences – the most obvious solution to have been proposed so far – would be prohibitively expensive. Similarly it is clear that, together with many drainage or 'managed retreat' schemes, such moves would also have a major and damaging impact on a number of important if marginal habitats, such as the region's salt-marshes, inter-tidal mudflats and sandbanks.

Spanning the Wash from a point just south of Skegness to Hunstanton 18 kilometres away, with an additional 5 kilometres at the Lincolnshire end to take it to higher ground, my own ark – the £2 billion Wash Tidal Barrier – therefore has many potential benefits. Benefits both to the region and the country as a whole, not just in terms of increased and more reliable defence against flooding but also improved environmental protection (and indeed enhancement), and the economic benefits which would flow from its being used as part of a viable and reliable source of 'green' and sustainable power.

The cost of this is unsurprisingly substantial. By providing timely and necessary protection for the people, property and infrastructure of this region of eastern England, however, the savings will more than outweigh the initial cost of building and then maintaining the Barrier. In any event the alternative – increasing the height of existing defences – would also be costly, since an estimated 200 kilometres of wall and bank would have to be raised for these to be effective, including all those along the region's many tidal rivers.

It is also significant that doing this would bring with it no other benefits beyond an improved level of flood protection. On the contrary: the sea beyond the new and improved defences would simply get deeper, thereby imperilling and eventually destroying the aforementioned marshland, mudflats and sandbanks and those species which live or overwinter on them. These same defences would also then be subject to even greater erosion by the sea, thereby increasing the maintenance costs year on year.

By contrast the Barrier would give us the opportunity to protect some of these more fragile environments, and to relocate and even improve the area's biodiversity by allowing new areas of habitat to develop, and by creating new ecosystems to host many of those species currently threatened or even displaced by climate change.

The creation of a lagoon behind the Barrier would also bring with it numerous economic benefits above and beyond the valuable injection of capital into the relatively weak economies of East Lincolnshire and West Norfolk. For example, the Wash currently provides very limited opportunities for commercial fishing whereas the new lagoon could be managed in such a way as to provide opportunities for both fish and seafood farming. The Barrier project also offers the opportunity to reclaim additional land, redeploying some of the existing mudflats and sandbanks for agricultural, commercial and leisure use.

The Wash, with its large tidal range, dangerous currents and high silt levels, is also at present very difficult to access for leisure. Once enclosed by the Barrier, however, and protecting popular resorts such as Hunstanton from the sea, the lagoon would be far better able to provide holidaymakers with a perfect environment for all manner of leisure pursuits, including sailing, rowing, angling, even swimming and diving. (A system of locks for smaller crafts would facilitate the development of new marinas, giving access to both the enclosed waters and the open sea.)

The final and not insubstantial benefit of the Barrier, and another advantage it enjoys over the more conventional option of raising existing defences, is its potential as a source of sustainable power. Most obviously this could be generated using the tidal power – with the potential to produce more than one gigawatt (equivalent to two nuclear power stations) – but also by siting wind turbines along the Barrier. With the lagoon functioning as a kind of storage battery, surplus power could be stored during periods of high wind and low demand and released at times of low wind and high demand.

To speed the design and completion of this important new ark, the Wash Tidal Barrier Corporation plc has now been established, a company created solely to promote and build the Barrier. The necessary team of people and companies is in place, contacts have been established with regional and national governments as well as other interested local parties and local landowners, and construction of the barrier should begin in 2011.

Printed in the United Kingdom
by Lightning Source UK Ltd.
131061UK00001B/70-519/P